THE ORDNANCE SURVEY GUIDE TO GUIDE TO HISTORIC HOUSES IN BRITAIN

Edited by

Peter Furtado, Candida Geddes,

Nathaniel Harris, Hazel Harrison and Paul Pettit

D0111497

W. W. NORTON & COMPANY

NEW YORK   LONDON

First published 1987 by
Ordnance Survey   and   Country Life Books
                         an imprint of
                         The Hamlyn Publishing Group Limited

First American Edition published 1987 by
W. W. Norton and Company, 500 Fifth Avenue, New York, N.Y. 10110

ISBN 0-393-30401-9

Maps Crown Copyright © 1987

Copyright © in text The Hamlyn Publishing Group Limited 1987

The contents of this book are believed correct at the time of publication; whilst every effort has been made to ensure that the factual information given is accurate, no liability can be accepted by the publishers for the consequences of any error. The representation on the maps of any road or track is not evidence of the existence of a public right of way.

Regional maps on pages 18-19, 72-73, 116-117, 145, 211, 251, 295 by Thames Cartographic Services, © The Hamlyn Publishing Group Limited and Ordnance Survey 1987.

Printed in Great Britain

Contents

How to use this guide

This guide is arranged regionally; a key map showing the area covered by each region is given on the opposite page. Within each region, the houses are listed alphabetically by county.

A detailed description is given of how to find the way to each house from the nearest town, together with the sheet number of the Ordnance Survey Landranger map which covers the area and the National Grid reference of the house site. Sections of Ordnance Survey maps are also included for over 50 of the houses.

The National Grid reference given for each house consists of two letters and four numbers. This pinpoints the location to within one kilometre on Ordnance Survey maps. The following example identifies how these references are constructed:

Audley End, Saffron Walden, Essex has a grid reference TL 5238.

TL   This identifies the 100 kilometre grid square in which the property lies and can be ignored from the point of view of locating the houses in this book.

52   Can be found in the top and bottom margins of the relevant Ordnance Survey Landranger map sheet (identified for each property). It is the reference number for one of the grid lines running North/South on the map.

38   Can be found in the left and right hand margins of the relevant Ordnance Survey Landranger map sheet. It is the reference number for one of the grid lines running East/West across the map.

These numbers together locate the bottom left hand corner of the one kilometre grid square in which Audley End appears.

When planning routes or driving to the areas of the properties included in this guide, the *Ordnance Survey Motoring Atlas of Great Britain* at 3 miles to 1 inch scale and the *Ordnance Survey Road Atlas of Great Britain* at 1:250 000 (approximately 4 miles to 1 inch) are ideal. For pinpointing the site once in the area use Ordnance Survey Landranger (1:50 000) or Pathfinder (1:25 000) maps. For London sites, the Ordnance Survey *ABC London Street Atlas* is a useful location guide.

Key to symbols

₽	Car park	❀	Plant sales
WC	Toilets	✎	Guide book on sale
⊖	Public transport	✗	Guided tours
♿	Access for the disabled	●	Restrictions on photography
⊟	Parties welcome	⋏	Playground
D	Dogs welcome	⋏	Nature trail
●	Refreshments available	NT	National Trust
⊓	Picnic area	EH	English Heritage
★	Free entry	WHM	Welsh Historic Monuments
◆	Shop		

M: Monday   T: Tuesday   W: Wednesday
Th: Thursday   F: Friday   S: Saturday   Su: Sunday

Map to regions

Scotland

HIGHLAND

GRAMPIAN

TAYSIDE

FIFE

CENTRAL

LOTHIAN

STRATHCLYDE

BORDERS

DUMFRIES
AND
GALLOWAY

NORTHUMBERLAND

1 TYNE AND WEAR
2 CLEVELAND
3 WEST YORKSHIRE
4 SOUTH YORKSHIRE
5 GREATER MANCHESTER
6 MERSEYSIDE
7 WEST MIDLANDS
8 BEDFORDSHIRE
9 BERKSHIRE
10 WEST GLAMORGAN
11 MID GLAMORGAN
12 SOUTH GLAMORGAN

DURHAM

CUMBRIA

The North

NORTH YORKSHIRE

LANCASHIRE

HUMBERSIDE

GWYNEDD

CLWYD

CHESHIRE

DERBY-
SHIRE

NOTTINGHAM-
SHIRE

LINCOLNSHIRE

STAFFORD-
SHIRE

LEICESTERSHIRE

SHROPSHIRE

**Wales and
Western Counties**

**Central
England**

NORFOLK

Eastern Counties

POWYS

HEREFORD
AND
WORCESTER

WARWICK-
SHIRE

NORTHAMPTON-
SHIRE

CAMBRIDGE-
SHIRE

SUFFOLK

DYFED

GLOUCESTER-
SHIRE

OXFORD-
SHIRE

BUCKINGHAM-
SHIRE

HERTFORD-
SHIRE

ESSEX

GWENT

GREATER
LONDON

**London and
Southern England**

AVON

WILTSHIRE

SURREY

KENT

SOMERSET

HAMPSHIRE

WEST
SUSSEX

EAST
SUSSEX

The West Country

DEVON

DORSET

ISLE OF
WIGHT

CORNWALL

Introduction

The buildings of Britain have been spared much of the havoc wrought by wars and revolutions in other countries. The relative security of British life has encouraged building with an eye to permanence and posterity; and in turn the stability and continuity of the social order ensured that much would survive despite individual whims and changes in taste and fashion.

At various times new owners have, of course, replaced or neglected buildings inherited from past generations. But more often British houses and castles have been modified or adapted over the centuries – enlarged in new styles, provided with the conveniences felt to be indispensable in a particular era, more or less redecorated inside, or given a face-lift outside by landscaping or the addition of the latest in the way of garden architecture or statuary. In such places, time has weathered and made a unity of the most apparently disparate elements. Houses have become part of the national heritage, preserving the past in stone, brick, timber, plaster and other materials. Naturally one can admire the end result with only the vaguest notions of history and architecture; but there is a richer pleasure in deciphering and relating the various human and material elements in a building.

The medieval heritage: Gothic

Time, troubles and the perishable nature of some materials have obliterated the houses and palaces built in Britain before medieval times; even the finest remains, such as the Roman palace of Fishbourne in West Sussex, are of archaeological interest rather than appealing directly to the visitor's eye as historic architecture. For a long period during the Middle Ages, too, the most imposing buildings were not domestic but military or ecclesiastical – castles built for Norman lords, and the churches, abbeys and cathedrals on which vast resources were expended during the 'Age of Faith'. Castle architecture is a separate and largely self-contained subject, but the great ecclesiastical styles are worth brief consideration as they had an influence on the details of medieval building, and in the 18th and 19th centuries enjoyed an extraordinary revival during which they were widely employed for both secular and ecclesiastical purposes.

After 1066 William the Conqueror and his followers introduced the style known in England as 'Norman'. This was in fact just the local version of European Romanesque, which was based on the use of round arches and massive supporting walls, pillars and buttresses – but which was nonetheless capable of creating a masterwork such as Durham Cathedral.

Gothic also originated abroad, in mid 12th-century France, and was imported towards the end of the century. By contrast with the massiveness of Romanesque, it is all line and light, soaring heavenward in a thin case of stone and glass. Whatever the change of spiritual outlook involved in this development, it was effected materially by engineering discoveries – involving pointed arches, rib vaulting and flying buttresses – that did away with the need for thick walls and massive pillars. Gothic in Britain is generally classified into three period styles: the simple Early English, the lavish and ornate Decorated, and the emphatically rectilinear Perpendicular Gothic, which was a distinctively English development without parallel on

The magnificent, beautifully proportioned 'double cube' room at Wilton House, Wiltshire.

the Continent. Decorative features also went through several phases – notably vaulting (fantastically ribbed and moulded in the otherwise austere Perpendicular style) and the long, slender and arched lancet window, which mutated into the decorative stonework designs known as tracery, variously geometric, riotously sinuous and strictly rectilinear. Vaulting, window tracery, crockets, finials, bosses and other Gothic features have had a long history of use in domestic architecture, and never more enthusiastically than in Victorian times.

The medieval hall
Whether in the castle or the manor house, medieval living arrangements long remained rudimentary. The most important area was the hall, the large common living- and eating-room in which the lord, his family and his retainers spent most of their time. It was a lofty place – its height ensured that smoke rising from the central hearth and lingering in the rafters would not stifle the inhabitants – with window openings that, for want of glass, were protected by wooden shutters. The lord and his family occupied a dais, raised a step or two above the rest of the hall, and had a separate private room adjacent to it; food and drink were brought in from the pantry and the buttery at the other end of the hall. Most early halls were built of wood and have therefore disappeared, but Oakham Castle in Leicestershire is a remarkable example of a 12th-century stone hall, aisled like a church because the timbers used for the roof were not long enough to span the entire width; the aisled sections have separate lean-to roofs. The timber roof nonetheless became one of the glories of English architecture, at its most elaborate and impressive in the 15th and 16th centuries. Extraordinary and decorative ingenuity was lavished on these roofs by English craftsmen, who filled not only halls but many otherwise insignificant churches with superb examples of their skills.

Many buildings, including manor houses, were timber-framed, or 'half-timbered'. Instead of stone walls carrying the weight of the structure, a sturdy framework of oak beams took the load. The spaces between the beams were usually packed with a dense, sticky mass of twigs caked with wet clay that had been mixed with straw or hair. The infilling (wattle and daub) was later often replaced by laths (flat strips of wood) or bricks. Finally the exterior would be plastered or covered with weatherboarding (overlapping boards). Half-timbered houses are widely but wrongly thought of as typically 'Tudor'; but that only reflects the fact that the great majority of surviving examples date from the 16th and 17th centuries. The technique is far older; an earlier surviving house such as Lower Brockhampton in Herefordshire (late 14th century) was certainly neither exceptional nor new in its own time.

In the late Middle Ages comfort became an increasingly important consideration; even castles were furnished with large windows and other amenities that were likely to reduce rather than increase their military effectiveness. Manor houses grew, though as yet in a somewhat ramshackle fashion, with one room after another being added on; frequently they formed a rectangle round a courtyard and, with the addition of a strong gatehouse, acquired a military air that was still felt to be a source of prestige. An important innovation was the use of bricks, initially imported from the Low Countries; they were not used on any scale until the 15th century, but then rapidly became very popular.

Palaces, prodigy houses and the Renaissance
The trend towards domesticity and comfort was affected surprisingly little by the Wars of the Roses; and when the new Tudor dynasty (1485-1603) brought

peace and relative stability, a long secular boom began. Ecclesiastical building came to an almost abrupt halt, and for 150 years English life was transformed by the erection of houses and mansions (and to a lesser extent palaces) in large numbers. Under the early Tudors the growing impulse towards display and comfort was not as yet refined by any underlying sense of style, and the visitor to a Tudor 'great house' is struck first of all by its enormous gatehouse, often adorned with coats of arms, and then by the fantastically jumbled skyline of twisted chimneys, vanes and turrets.

The hall was still the central feature of the house, but the number and size of the other rooms grew steadily. A new and distinctively English feature, the long gallery, appeared on the first floor – a room spacious and comfortable enough for the proprietorial family to enjoy spending most of their indoor leisure time in it, away from their retainers and dependants; as the name implies, it was long enough to be strolled about in. Walls were generally panelled in wood, which was often carved with the distinctive linenfold design, tremendously popular in Tudor times. (It was exactly what the name indicates: a stylised image of a piece of folded linen.) Otherwise, such style as there was derived mainly from Perpendicular Gothic: window tracery and stained glass; oriel windows (bay windows above ground-floor level); and depressed arches in gatehouse entrances and windows, and above fireplaces. The proper fireplace, equipped with a flue leading to a chimney, was perhaps the aspect of 'the general amendment of lodging' in early Tudor times that impressed contemporaries most; the insistent corkscrew shape of the stacks was no doubt the owner's way of advertising that his house was right up to date.

The first significant English contacts with Renaissance art took place during the reign of Henry VIII (1509-47). The King himself imported an Italian sculptor to carve the tomb of his father, Henry VII, in Westminster Abbey. His chief minister, Cardinal Wolsey, built Hampton Court Palace, but in a style that revealed the superficiality of English feeling for Renaissance architecture: the structure, including Henry's great hall, remained tradition- ally Gothic, and the Renaissance elements were confined to aspects of the decoration, such as the medallions with terracotta busts of Roman emperors on the gateways.

Renaissance influences made only slow progress in England, partly because Henry VIII's break with the Catholic Church virtually eliminated direct Anglo-Italian contacts. (It also led to the Dissolution of the Monasteries, whose extensive lands were acquired by the upper classes; the origin of stately homes such as Newstead Abbey and Woburn Abbey remains apparent in their names.) After about 1536 English notions of Renaissance art and architecture were derived from French and Flemish sources, which inevitably distorted them in transmission. Even after the publication of the first English book describing the classical Orders, John Shute's *First and Chief Groundes of Architecture* (1563), English taste remained eccentrically provincial and eclectic, and during Queen Elizabeth's reign (1558-1603) created a number of striking and unusual buildings that have become increasingly admired in recent years.

Under Elizabeth the upper and middle ranks of English society prospered greatly, and despite the crises of the reign – by no means over even when the Spanish Armada had been defeated – the dominant classes were full of self-confidence and almost maniacally self-assertive. Their taste for competitive display helped to sustain a long surge of building and rebuilding which reached extravagant heights in the massive showpieces – the 'prodigy houses' – commissioned by the great men of the realm. Even Elizabeth's sober chief minister, William Cecil, Lord Burghley, shared the prodigy house mania, building both Burghley House in Northamptonshire and Theobalds in Hertfordshire (since demolished).

During this period the influence of the Italian Renaissance began to make itself felt in Britain. The Renaissance ('rebirth') was a great cultural and artistic movement that conceived of itself as breaking with the medieval past and returning to the values and practices of ancient Greece and Rome. In architecture, the abundance of Roman remains in Italy and the wide circulation of a Roman treatise by Vitruvius enabled Alberti, Brunelleschi, Palladio and others to devise an Italian Renaissance style adapted to contemporary needs but always 'classical' (that is, Greco-Roman) in general intention, and certainly so in its details. In the hands of the Italian masters, Renaissance architecture became the expression of a supremely calm, balanced and harmonious spirit that was to remain associated with subsequent European 'classicism' in many of its forms; but in Britain and elsewhere the classical tradition was also to prove surprisingly adaptable to changing moods and requirements, a fact that accounts for its survival over the centuries.

The most famous of the prodigy houses are associated with Robert Smythson (c. 1536-1614), an architect of very great ability though his status remained that of a mason or 'surveyor'. His patrons undoubtedly made many of the critical decisions about the larger design features of their houses, and this may account for the otherwise curious fact that Smythson's first major commission, Longleat in Wiltshire, is the most classically compact of his buildings. In the next, Wollaton Hall, Nottinghamshire, the corners have been drawn up and out to make towers, indicating that Englishmen had not yet abandoned military-style fantasy – even though they were also submitting to the foreign (Renaissance) fashion for unmistakably civilian balustrades and colonnades.

This tendency is even more marked at Hardwick Hall in Derbyshire, where the three-storey main block is surrounded by six four-storey towers. Hardwick is probably also Smythson's work, and again combines Renaissance features such as the colonnaded façade and balustraded skyline with the Elizabethan taste for a weighty, vaguely castellated general appearance embellished by bays and quantities of glass. The well-known jingle 'Hardwick Hall/ More glass than wall' conveys some of the pride and wonder evoked by vistas of this material, whose lavish use had not long before been an ecclesiastical monopoly. The interior of Hardwick exemplifies the changed function of the hall, now placed neatly on the axis of the main entrance to form a grand reception area, while separate state and private rooms on different floors catered for the public and personal sides of the owner's life.

In Elizabethan houses ornament ran riot, suggesting that people were positively uncomfortable when forced to contemplate unfilled spaces. Fabrics were densely embroidered, and walls were hung with tapestries and portraits; woodwork was carved and plasterwork moulded into a wide variety of interlacing, geometric and floral designs.

The building boom went on into the reign of the first Stuart king of England, James I (1603-25). James's prodigality and chronic shortage of money led him to bestow or sell titles on an unprecedented scale, further encouraging his subjects to build dwellings appropriate to their improved status. The first great Jacobean house, however, was Knole in Kent, remodelled by Queen Elizabeth's cousin Thomas Sackville, Earl of Dorset; the outside is characteristically English and eclectic, combining a colonnade in a minor Order, the Tuscan, with carved Dutch gables, which became a very popular feature of Jacobean buildings. The greatest Jacobean prodigy house that has survived intact is Hatfield House in Hertfordshire, built in brick by Robert Cecil, the son of Elizabeth's chief minister. Cecil had been confirmed in power and created Earl of Salisbury by James I, to whom he had felt obliged to present the palatial Theobalds; in return he received the one-time ecclesiastical palace of Hatfield, which he plundered for building materials to use in his new prodigy house. The episode is characteristic of the way in which the new country-house England supplanted the medieval order.

Inigo Jones and 17th-century classicism

The first self-proclaimed English architect and exponent of pure classicism appeared on the scene quite suddenly. Inigo Jones (1573-1652) was the son of a Smithfield clothworker. Almost nothing is known of his early life except that as a young man he somehow managed to visit Italy and Denmark. In 1605, already in his thirties, he began his career at Court as an omni-competent designer, notably of costumes and sets for masques – pageant-like performances, combining poetry, music and spectacle, in which the entire Court took part. In 1613-14 he revisited Italy in the train of the Earl of Arundel, studying antique architecture with the help of the Paduan architect Palladio's book *I Quattro Libri dell'Architettura* (1570); Jones's copy, containing his notes, still survives. As both a writer and practitioner of domestic architecture in a superbly harmonious classical style, Palladio was to have an enormous influence on English architecture. Jones became the first English Palladian after he was appointed Surveyor of the Works, in charge of all royal building, in 1615. His principal surviving works are the Queen's House at Greenwich (now the National Maritime Museum) and the Banqueting House at Whitehall; he is less certainly associated with several country houses, including Wilton House in Wiltshire.

After the heavy self-assertion and idiosyncracies of the prodigy houses, the cool, elegant simplicity of Jones's buildings comes as a relief and a surprise; it is hard to believe that James I's courtiers were not astonished when they saw the Banqueting House take shape in 1619-22. It revealed that the Renaissance style was not just a glorious repertoire of decorative devices – as English builders had assumed on the basis of second-hand information – but an intellectually coherent system of relationships that determined the proportions of all the component parts.

The simplicity of Jones's designs is therefore deceptive. Instead of the far-flung E- and H-plans of Elizabethan and Jacobean houses, the Queen's House and the Banqueting House are compact rectangles with flat balustrad-ed roofs, neatly grouped chimneys, and closely integrated, regular decorative details including columns, pilasters, swags (leafy chains) and pedimented windows. Their interiors, too, demonstrated that the mass of ornament which had increasingly overwhelmed the English house was not necessary to create an effect of grandeur.

The English Civil War effectively ended Jones's career. But his assistant John Webb (1611-72) and a highly talented gentleman-amateur, Sir Roger Pratt (1620-85), went on to apply classical principles to the building of country houses; at the Vyne, in Hampshire, Webb introduced the temple-style portico, with columns supporting a great pediment, which was to become one of the most popular features of English country houses in the classical tradition.

Another influential figure was Hugh May (1621-84), who had shared Charles II's exile in Holland and brought back with him a knowledge of the Dutch version of Palladianism. This entailed the use of brick, a material that the English understood very well, and in time evolved into a type of building popular for generations because its plain, quiet dignity so admirably suited well-to-do people without any aristocratic pretensions. Its standard features included various classical elements that were used only decoratively, notably a large central pediment on the main façade; a hipped (four-sided) roof that, by dispensing with gables, allowed the placing of a characteristically elegant, unbroken cornice along the line of the eaves; and dormer windows lighting a top storey placed within the roof to avoid spoiling the proportions of the building. In this and other types of house, the sash window – in its day an ingeniously designed convenience – became commonplace from about 1700.

Although the type of building just described has often been called a 'Wren house' (when it has not been labelled 'Queen Anne'), it has no significant connection with Sir Christopher Wren (1632-1723). It is in fact ironic that the greatest British architect should warrant only the briefest mention here, as most of the famous Wren buildings are not houses. However, the Fountain Court at Hampton Court Palace is his work, and so is Flamsteed House (the Royal Observatory, Greenwich Park), built in 1675 for the first Astronomer Royal, John Flamsteed – an appropriate conjunction since Wren himself had been a professor of astronomy before turning to architecture.

Wren's extraordinary genius, and the sheer scale and number of his buildings, established classicism as *the* English style for all imaginable purposes. And at almost the same time, Scottish architecture also succumbed to classicism after a final splendid flowering of castle-building in what the Victorians named 'Scottish baronial' style. It was only after the Restoration of 1660 that Scotland was seriously invaded by classicism of the sort favoured by the English. As surveyor-general for Scotland, Sir William Bruce (d. 1710) introduced a thoroughgoing classicism in his work on Holyrood house, Edinburgh, and in his own Kinross House, Tayside; and this supplanted the native style. Scottish architecture became part of the main British tradition, with at least one favourable result, in that Scotland produced a series of major architects who were able to take full advantage of the growing wealth and sophistication of British society.

The English Baroque

For most of the 17th century Italy remained the dynamic centre of European art. One of its most astonishing creations was the Baroque style, which united architecture, painting and sculpture in a vast, all-embracing celebration of grandeur and glory – above all, in Italy, the grandeur and glory of the Catholic Church. The supreme exponents of Baroque, the Italians Bernini and Borromini, worked on the largest possible scale to create a heroic art in which sweeping, curvaceous forms were organised with a sculptor's sense of the effective interplay between masses and voids. These fluent forms, and the wealth of carved and moulded ornament, blurred the lines between sculpture and architecture; and painting, too, was recruited into the service of drama and illusion: the Baroque painter – especially the ceiling painter – carries the spectator's eyes straight out of the house and up into the heavens.

As it travelled north-west, the Baroque was put into the service of new patrons. At Versailles it glorified the absolute monarchy of Louis XIV, the Sun King; while in Britain this ultra-Catholic and absolutist style was taken up by the great Whig and Protestant landowners who had increasingly become the country's real masters. Chatsworth, Castle Howard and Blenheim are grand monuments to the generation of Whig grandees who saw through the long wars against France and established the Hanoverians on the throne, to their own considerable political advantage.

For all its extravagance and complexity, Baroque was a development of Renaissance classicism – a wayward one in which the dignified classical vocabulary is manipulated and distorted in the interests of a theatrically

12

heroic emotion. English architects never fully embraced the ecstatic curvi-linearity of full-blooded Continental Baroque, and the work of Wren and his successors is sometimes described by the compromise term 'Baroque classicism', indicating the continuing presence in it of a certain restraint or a temperamental preference for the foursquare. All the same, English Baroque proved a grandiose although brief episode. Chatsworth in Derbyshire was remodelled for the Duke of Devonshire by William Talman (1650-1719) in 1687-96. Then, in 1699, Sir John Vanbrugh (1664-1726), soldier of fortune and fashionable playwright, turned architect and designed Castle Howard in North Yorkshire for the Earl of Carlisle. In 1705, while this huge house was still under construction, Vanbrugh won the commission for Blenheim Palace, which was built at the national expense as a home for (and monument to) the Duke of Marlborough, who had led the victorious Allied armies against Louis XIV. At Castle Howard, and even more so at Blenheim, Vanbrugh was able to indulge his taste for the colossal, creating buildings that in size and complex interaction of parts had no parallel in earlier British architecture. Although individually 'classical', his giant orders, far-flung wings, tall domes, towers, and assorted urns and sculpted figures on the skyline are a world away from the cool, compact style of Palladio and his disciple Inigo Jones.

Vanbrugh was assisted by Nicholas Hawksmoor (1661-1736), who had previously spent twenty years working for Wren; quite why this gifted architect allowed himself to be overshadowed in this fashion is not clear. Apart from his famous City churches, Hawksmoor designed the striking Mausoleum at Castle Howard. Other Baroque architects of distinction were Thomas Archer (c. 1668-1743), who worked at Chatsworth, and James Gibbs (1682-1754), whose career was impeded by his Catholic and Tory affiliations; as well as such famous buildings as St Martin-in-the-Fields in Trafalgar Square, London, and the Radcliffe Camera at Oxford, he designed Orleans House, Twickenham, and created a very late Baroque hall at Ragley, Warwickshire (begun c. 1750).

The complex, ornate interiors of Baroque houses called on a wide range of native skills. However, for the illusionistic decorative painting characteristic of Baroque interiors, foreigners – notably the team of Verrio and Laguerre – were imported. One British painter of some distinction emerged in Sir James Thornhill (1675-1734), who worked at Blenheim and elsewhere, and was also the architect of Moor Park, Hertfordshire.

The Palladian revival

Early in the 18th century one of the most extraordinary revivals in the history of architecture took place: Baroque was vigorously attacked and overthrown even before Vanbrugh's two major projects had been completed. Instead, strict adherence to the classicism of Inigo Jones and Palladio became the order of the day. The publication in 1715 of two books, Colen Campbell's *Vitruvius Britannicus* and a new translation of Palladio, began the new trend. Campbell (d. 1729) converted the wealthy and cultured Lord Burlington (1694-1753), who dismissed Gibbs and gave the previously obscure Scottish architect the job of finishing his London mansion, Burlington House. Campbell went on to design the most slavish of all tributes to Palladio, Mereworth Castle in Kent, which is a virtual copy of the Italian master's famous Rotunda. He worked on the grand scale at Houghton Hall, Norfolk, built for the Prime Minister, Sir Robert Walpole, and more modestly at Stourhead, Wiltshire, and elsewhere.

Meanwhile, Campbell's patron, Lord Burlington, made himself into an architect of some distinction. After taking on Campbell, he studied Palladio at first hand in Italy (1719-20) and then returned to Britain and became the high priest of the new movement. His wealth, influence and puritanical insistence on 'correctness' (as laid down by Palladio) combined to make the Palladian country house the norm for an entire generation. Burlington's best surviving works are his own villa, Chiswick House, and the Assembly Rooms, York.

Burlington brought back with him from Italy a curious, only semi-literate artist named William Kent (*c.* 1685-1748), who had seriously mistaken his vocation. Kent believed he was a good history painter in the grand manner – but he was not; and even Burlington's powerful influence failed to forward a career punctuated with rebuffs and humiliations. But Kent *was* a designer of genius. In his forties he became an architect (Holkham Hall, Norfolk), but he was more original as an interior designer – if only because almost nothing was known about ancient Roman interiors, leaving the gifted designer free to combine and permutate antique elements in accordance with his own tastes. Kent also initiated the 18th-century revolution in landscape gardening; he and his successors, Lancelot 'Capability' Brown (1716-83) and Humphrey Repton (1752-1818), swept away the formal 'French' garden, and created the 'English' garden, in which – after careful preparation – 'nature' apparently ruled supreme.

As towns became centres of fashionable, professional and commercial life, the need for new forms of housing became apparent. The problem arose relatively early at Bath, which was the first example of a new phenomenon; the fashionable spa and resort town, thronged with visitors. The response was the earliest of the great town-planning schemes characteristic of the Georgian period (others were the New Town, Edinburgh, and Regent's Park, London). From 1729 John Wood the elder (1704-54) built Queen Square and the Circus, finished by his son John Wood the younger (1728-81), who went on to build the magnificent Royal Crescent (1761-65). The Woods' inspired adaptation of classicism to substantial terrace houses, conceived on the grand scale as collective units, provided a model for generations of British builders.

Neo-Classicism

With the passing of Burlington's generation of Palladians, a number of different tendencies appeared in British classicism. Although no fundamental change was involved, the label 'Neo-Classical' tends to be used of art and architecture produced in this tradition after about 1750 – partly, at least, to accommodate it to the genuinely new Neo-Classical movement developing on the Continent, which did entail a significant break with the immediate past.

One advantage possessed by the new generation was the ability to study antiquity at first hand, rather than through the writings of Palladio and other Renaissance architects. Wealthy enthusiasts for the antique even made it possible for James Stuart and Nicholas Revett to make the first accurate on-site measurements and drawings of the masterpieces of Greek architecture, and their book *The Antiquities of Athens* (1762, 1789) revealed the austere simplicity of the Greek style to readers whose notions of classicism were based entirely on Roman and Renaissance buildings. However, although Stuart designed a thorough-going Greek temple at Hagley (West Midlands), the specifically Greek Revival did not begin until the 19th century, and when it did was essentially an institutional style for banks, insurance companies and the like.

From our point of view, the major figure of this generation was the Scottish architect Robert Adam (1728-92), who worked at many places described in this book – Syon House, Osterley Park and Kenwood, all in or near London;

Nostell Priory and Harewood House in West Yorkshire; Kedleston Hall, Derbyshire; and Saltram House, Devon. In most cases his task was to remodel an existing building or finish off another architect's work, for by this time most of the great 18th-century houses had already been built or were under construction.

Nonetheless, all Adam's best houses carry his individual stamp. His exteriors are Palladian, but with a more sculpturesque feeling than Campbell's or Burlington's; Adam admitted to being influenced by Vanbrugh and wrote that he regarded 'movement' as his most important contribution to architecture. However, Adam's interiors are undoubtedly his greatest achievements – dazzling displays of polished marble, gilding, beautifully moulded plasterwork, bronze, glass, wood, etc., combining lightness with sumptuousness. Like Stuart, Adam was a scholar; he had published measured drawings of Diocletian's Palace at Spalato (now Split, Yugoslavia), studied the interiors of newly excavated Roman houses at Pompeii, and drew on a variety of Renaissance and supposedly 'Etruscan' (actually Greek) sources in his decorative work. The fusion of these influences is a unique feat, made possible by the fact that Adam was the first British interior designer in the modern sense, meticulously overseeing every detail of the work in hand, whatever the medium.

Castles, pagodas and the Picturesque

There were other sides to Adam's activities. In his native Scotland he worked on a number of houses such as Mellerstain, in which he adapted himself to the old 'Scottish baronial' castle style. In England, for the dilettante and connoisseur Horace Walpole he designed the round tower at Strawberry Hill, Twickenham. This small 'castle', created by Walpole and his friends over a number of years, was an early example of the 'Gothick' – a revival of the medieval style that, while not necessarily unscholarly, was certainly playful rather than antiquarian in spirit; some years earlier Walpole had published a hugely successful Gothick novel, *The Castle of Otranto*, which also exploited the appeal of the 'barbaric' past to the ultra-civilised sensibilities of his contemporaries.

The Gothick was one of several manifestations of discontent with 18th-century classicism and sweet reasonableness; these did not overthrow Neo-Classicism but coexisted with it, and sometimes modified its impact. Chinoiserie, the cult of things Chinese (or mock-Chinese), made its greatest impact on interior decoration, but a respectable Neo-Classical architect, Sir William Chambers, designed the pagoda at Kew and a Chinese dairy at Woburn Abbey. Much more important, and closely related to Gothick, was the theory of 'the Picturesque', which emphasised the distinctly unclassical virtues of irregularity and ruggedness, making a cult of rocks and ruins that transformed many country house landscapes, filling them with brand-new crumbling abbeys and similar 'follies'. One of the champions of the Picturesque, Richard Payne Knight, built himself a massive castle, which gave a great impetus to the medievalising vogue; once established, the taste for castles or at least the 'castle look' persisted throughout the 19th century.

Confronted with a choice of styles, some architects – notably Henry Holland (1745-1806) – remained faithful to Neo-Classicism. Most, however, proved willing to adapt, following the example of the prolific James Wyatt (1747-1813). He designed the classically elegant Heaton House, Greater Manchester; created a number of Adam-style interiors; built 'Gothic' castles of varying eccentricity; and is rather unfairly remembered for the most fantastic of all follies, built for the 'oriental-Gothick' writer William Beckford – Fonthill Abbey, whose 225-foot tower collapsed in 1825.

Wyatt's near-contemporary John Nash (1753-1835) was an equally versatile designer of country houses, but his career reached its climax a generation later, during the Regency period. He was taken up by the Prince Regent (later George IV), and devised the last of the great Georgian achievements in town planning, the Regent's Park and Regent Street Development in London. In an utterly different vein, he built the extra-ordinary but delightful Royal Pavilion, Brighton, in fantasy Indian style as a holiday home for his master. He also began Buckingham Palace, but was abruptly forced into retirement after the death of George IV in 1830.

The other great architect of the period was Sir John Soane (1753-1837), famous during his lifetime for his work on the Bank of England. He manipulated the Neo-Classical style in a highly original manner to produce romantic and mysterious effects. This is now best seen at his own house (Sir John Soane's Museum) in Lincoln's Inn Fields, London; its complicated layout, changing floor-levels, top lighting and multitude of mirrors create a striking, enclosed but surprisingly comfortable atmosphere.

The Victorians

The Victorian age (1837-1901) transformed Britain. This was the age of rapid industrialisation, huge population growth, cities, factories, railways, world-wide commerce and far-flung empire. Masterful engineers tunnelled through the Pennines, bridged the Forth and raised vast structures of metal and sheet glass such as the Crystal Palace and the great railway stations. Buildings of all sorts were needed and provided in quantities – slums and suburbs, churches and clubs, factories, mills, warehouses, banks, big business and civic buildings. And, of course, houses for the old and new rich, including enormous country houses and castles.

One of the striking things about Victorian architecture is the absence of any dominant style. Styles had overlapped or coexisted in earlier periods, but there had never before been anything like the Victorians' eclecticism – their willingness to build in any manner, whether ancient, medieval or modern, native, foreign or sheerly exotic. No new, distinctively 'Victorian' style emerged from this profusion: the age was one of successive or simultaneous revivals. The style with the strongest claim to high seriousness was Gothic, which thanks to the influence of A. W. N. Pugin (1812-52) ceased to be associated with the picturesque and mysterious; in the hands of such eminent High Victorians as William Butterfield (1814-1900) and Sir George Gilbert Scott (1811-78) it was used with scholarship and vigour for ecclesiastical building – and also for the Houses of Parliament and St Pancras railway station.

By contrast, Neo-Classicism ceased to be high fashion, though it was still found appropriate for many civic buildings (especially in the new cities and towns created by industrialism) and for solid middle-class terraces. The Victorians' taste for mass and ornament was better served by the neo-Renaissance *palazzo* style, with its powerful arches and rusticated surfaces (that is, surfaces in which the masonry consists of blocks with chamfered edges, creating an emphatic pattern of blocks and grooves). The major pioneer of the style was Sir Charles Barry (1795-1860), who quite literally acclimatised it to Britain by roofing over the Italianate open courtyard, which became a grand balconied hall. Followers of the style included Albert, the Prince Consort, who designed the royal residence, Obsorne House, on the Isle of Wight.

There were many more styles; and also sub-styles for the Victorian blend of eclecticism and scholarship encouraged architects to achieve a spurious originality by working in some national or period version of a style – for example, Dutch or French, rather than Italian, Renaissance. This proliferation was nowhere more marked than in country houses, where size, spectacle and instant historic effect were at a premium – notoriously so with vulgar newly rich magnates and financiers who were setting themselves up as 'lords of the manor', but also in fact with many supposedly unvulgar aristocrats who had managed to benefit from the booming Victorian economy.

The size of Victorian country houses was partly the result of a new pattern of weekend entertaining, made possible by the railways, which could bring or take away hordes of visitors within a few hours; these had to be accommodated, and so did the hordes of servants who looked after them. New rooms – for example the smoking room and gun room – gave guests the opportunity to spread out, and also catered for the Victorian obsession with specialisation, even more apparent in the complex arrangements below stairs. The Victorian country house has plausibly been likened to a rabbit warren, and the achievement of architects in making it work effectively should not be underestimated.

As accomplished professionals, most architects produced designs in whatever style their clients required. Barry adopted his preferred Renaissance style for Cliveden in Buckinghamshire, but also enlarged Dunrobin Castle in Scottish baronial – and rebuilt the Houses of Parliament in Gothic. Joseph Paxton (1801-65), designer of the severely functional iron-and-glass Crystal Palace, had no qualms about building Mentmore Towers, Buckinghamshire, in the Elizabethan manner of Wollaton. Anthony Salvin (1799-1881) was a prolific designer of Gothic, Elizabethan and Jacobean mansions. One architect with firm convictions was William Burges (1827-81), who lived in a private Gothic fantasy-world which, finding the ideal patron, he expressed lavishly at Cardiff Castle and Castell Coch in South Glamorgan. But sometimes no British architect would do: to build Waddesdon Manor, Buckinghamshire, in the style of a Loire château, Baron de Rothschild thought it necessary to import a bona fide Frenchman in the person of Gabriel-Hippolyte Destailleur!

Into the 20th century
Eventually a reaction set in against Victorian revivalism, especially in more modest domestic building. Pugin, John Ruskin, William Morris and the Arts and Crafts movement all contributed to a new emphasis on honest craftsmanship, the use of local and appropriate materials, and the avoidance of period styles – or at any rate a restrained and tasteful use of them. Philip Webb (1831-1915) was the first significant architect of the movement, and the Red House he built in 1859 for Morris at Bexleyheath was its first monument. Houses in this tradition were generally informed and unpretentious, horizontal in emphasis, respectful of local materials and associations, and 'Old English' in feeling, with a more or less definite medieval manor house atmosphere. Richard Norman Shaw (1831-1912) was influenced by this tradition, for example at Adcote in Shropshire, though he worked in other manners including his own 'Queen Anne' version of the 17th-century town house.

C. F. A. Voysey (1857-1941) took the 'Old English' manner into the 20th century, as did Sir Edward Lutyens (1869-1944). However, the spectacularly successful Lutyens later turned to a grandiose Neo-Classicism which reached a literally imperial apotheosis with the Viceroy's House, New Delhi. He is an appropriate figure with whom to end this outline, as his immense granite Castle Drogo (1910-30) was the very last country seat in the grand manner.

1 Fenton House
2 Dickens House Museum
3 Wesley's House
4 Sir John Soane Museum
5 Dr Johnson's House
6 Marlborough House
7 Lancaster House
8 Apsley House
9 Kensington Palace
10 Linley Sambourne House
11 Hogarth's House
12 Carlyle's House
13 Banqueting House
14 Kew Palace
15 Syon House
16 Marble Hill House
17 Ham House
18 Squerryes Court

WARWICKSHIRE

NORTHAMPTON-SHIRE

BEDFORDSHIRE

Luton

GLOUCS

BUCKINGHAM-SHIRE

HER

Oxford

OXFORDSHIRE

Swindon

G

Bos
Mar

Osterley Park House

Orleans House

15

Basildon
Park

BERKSHIRE

Reading

Swallowfield
Park

Hampton
Court

Claremont House

Stratfield
Saye House

The Vyne

Basing House

Guildford
House

Hatchlands

Clandon
Park

Poles
Lac

WILTSHIRE

Basingstoke

Guildford

Loseley House

SUR

Albury
Park

HAMPSHIRE

Jane Austen's
House

WEST

Avington Park

Alresford
House

Mottisfont
Abbey

Salisbury

Bishop's
Waltham
Palace

Petworth House

SUSSEX

Breamore
House

Broadlands

Southampton

Uppark

Parham Park

DORSET

Titchfield Abbey

Portsmouth

Goodwood House

Palace House
Beaulieu

Bournemouth

Norris
Castle

Osborne
House

Nunwell
House

Arreton Manor

Morton Manor

Haseley
Manor

ISLE OF
WIGHT

Appuldurcombe House

London and Southern England

Ipswich

FORDSHIRE

ESSEX

Harlow

Southend

Forty Hall

EATER

Kenwood House
Keats House
1
9 8 4 2 3
10 7 6 5
iswick 12
use 11 13
Chiswick Ranger's
House House
Southside
House
Little Holland House
Whitehall

Queen Anne's
House

Eltham
Palace

Down
House

Gad's Hill
Place

Owletts
Cobham
Hall

Temple
Manor

Tudor House,
Margate

Quex Park

Eastbridge
Hospital

Canterbury

Quebec
House

Detillens
Knole
18
Chartwell

Ightham
Mote

Stoneacre

Maidstone

Boughton
Monchelsea
Place

Dover

Riverhill
House

Penshurst
Place

KENT

Godinton Park

Sackville
College

rawley

Standen

Finchcocks

Pattyndenne

Port Lympne

Great
Maythan
Hall

Smallhythe Place

Sheffield
Park

Haremere
Hall

Legh
Manor

Bateman's

Great
Dixter

Brickwall
House

Beeches
Farm

Lamb House

Danny

SUSSEX

Hastings

ewtimber
Place

reston
Manor

Anne of
Cleves
House

Glynde
Place

Brighton

Firle
Place

Michelham Priory

yal Pavilion,
Brighton

Monk's
House

Clergy House,
Alfriston

Palace House, Beaulieu

Brockenhurst, Hampshire

Most people know Beaulieu as the home of the National Motor Museum, and indeed it is quite possible to miss Palace House and the ruins of the old abbey standing some little distance away. The house is the home of Lord and Lady Montagu, and has been in the family since 1538, when it was acquired after the Dissolution of the Monasteries. The building is a slightly odd combination of three styles: it was begun in the 14th century as the gatehouse to the abbey (the outline of the main entrance, on the south side, can still be seen); modified in the 1730s by the 2nd Duke; and in the 1870s converted and extended as a 'Scottish baronial' mansion by the architect Andrew Blomfield for Henry, 1st Baron Montagu of Beaulieu. The house is very much a family home, and thus only some rooms, those in the old gatehouse, are open to the public. The central arch of the original great gatehouse, in today's lower drawing room, is now filled by a very imposing Victorian fireplace. Throughout the house the decorations are mainly Victorian, and there is some good furniture and interesting family portraits. A novel and unusual feature of the house is that each of the rooms is peopled with a tableau of wax figures in period costume representing members of the family from earlier generations.

☎ Beaulieu (0590) 612345

10 m E of Brockenhurst on B3055

SU 3902 (OS 196)

Open May to Sept daily 1000-1800, Oct to Apr (exc 25 Dec) 1000-1700

P WC (limited access) D (seasonal) Monorail, veteran bus ride

Breamore House

Breamore, Hampshire

Built of rose-red brick with stone facings, this large Elizabethan manor house was completed in 1583. The Dodingtons, who owned it, were an unfortunate family. On April 11th, 1600 William, in a state of anxiety over a lawsuit, threw himself from the steeple of a London church in broad daylight. His son William, although knighted by James I, fared little better: in 1629 his wife was murdered by their son Henry in Breamore House. Henry was hanged in Winchester jail a year later, though legend claims he was 'hanged within sight of the house in which he was so untimely born'. The house now belongs to descendants of Sir Edward Hulse, who bought it in 1748. The visitor is admitted to the main rooms on the ground floor and to the east wing, where there are Tudor bedrooms and an Elizabethan four-poster bed. A fire in 1856 destroyed much of the original decoration, and much of it dates from the subsequent rebuilding, but the furniture survived. On the first-floor landing hangs a very rare English pile carpet, and at the top of the staircase there is an extraordinary set of fourteen paintings, each showing a different kind of mixed-race marriage possible in 17th-century Mexico. They were painted by an illegitimate son of Murillo, who had a studio in Mexico. The kitchen is interesting, as it was used until quite recently and is complete with all its fittings.

☎ Downton (0725) 22468

12 m S of Salisbury on A338 turn W to North Street and Upper Street

SU 1519 (OS 184)

Open April T, W, Su and Bank Hols; May, July and Sept also Th, S; daily in August 1400-1730

⊖ (½ m) Ⓟ WC ♿ 🚻 D (grounds only)
♣ 🍴 🎠 ◆ ✿ 🐕 ● (not in house)

Broadlands

Romsey, Hampshire

The elegant Palladian mansion of Broadlands was the home of Lord Mountbatten, and is now that of his grandson Lord Romsey. A previous owner, and the man responsible for transforming it from an ordinary 16th-century house into the fine manor we see today, was the 2nd Viscount Palmerston, who in 1766 employed Capability Brown to landscape the grounds and also make improvements to the house. Brown gave the house a grand new portico and refaced it in the fashionable yellow-grey brick, and in the 1780s his protégé and son-in-law Henry Holland added the east front portico and domed hall. This domed entrance hall leads to the sculpture gallery, both rooms containing parts of Lord Palmerston's collection of antique and 18th-century sculpture. The main ground-floor rooms are decorated in the Adam style. Most of the decorative plasterwork was done by Adam's favourite plasterer, Joseph Rose, that in the saloon being particularly fine. There are four paintings by Van Dyck in the dining room; 18th- and 19th-century portraits in the drawing room, and a collection of Wedgwood. Broadlands has had a great many distinguished visitors during its lifetime, including royalty, and the house has many historical associations. One room is devoted to the 3rd Viscount Palmerston, Britain's popular mid-19th-century prime minister, and contains the desk at which he wrote standing up. There is a display of model warships commanded by Lord Mountbatten, and a permanent Mountbatten exhibition in the 17th-century stables.

☎ Romsey (0794) 516878

S of Romsey off A31

SU 3520 (OS 185)

Open Apr to end July daily exc M (but inc Bank Hol M); Aug, Sept daily 1000-1800

⊖ 🅿 WC ⽧ 🈺 ♣ 🍴 🎠 ◆ ⚹ ● (no tripods)

Jane Austen's House

Chawton, Alton, Hampshire

The 300-year-old house, now the Jane Austen Museum, was the author's last home, where she lived with her mother and her sister Cassandra from 1809 until her early death in 1817. Little is known of the house's early history, except that it was built as an inn, and remained so until occupied by Jane's third brother Edward, who offered it to his mother when the Rev. George Austen died. The outside of the house looks very much as it did in Jane's time, but the interior was considerably altered after Cassandra's death in 1845, when it was divided into three separate dwellings for farm workers. However, it has now been restored as far as possible, and visitors can see the drawing room, vestibule, dining room and upstairs parlour. Jane, who wrote *Mansfield Park, Emma* and *Persuasion* here, used the dining room for her writing, and a creaking door warned her when she was about to be interrupted. There is only one piece of her own furniture in the house, a small round table which she is believed to have used for writing, but there are a few personal items including a lock of hair, some jewellery and a patchwork quilt made by her and her mother. She was very fond of embroidering: her younger sister described her as 'a great adept at overcast and satin stitch, the peculiar delight of that day'. The particular feature of the museum is its large collection of documentary material, including many letters written by Jane, and there are also illustrations of her books and of her many other homes.

☎ Alton (0420) 83262
1 m SW of Alton off Alton by-pass in Chawton village
SU 7037 (OS 186)

Open Apr to end Oct daily; Nov, Dec and Mar W-Su; Jan and Feb S and Su 1100-1630

⊖ 🅿 WC ♿ (limited access) 🍴 ♦ ☂ ❖ ✻

Stratfield Saye House

Stratfield Saye, near Reading, Hampshire

Stratfield Saye, like Blenheim, was acquired with public money for a national hero, in this case the Duke of Wellington, but there the resemblance ends; Stratfield Saye, originally built between 1630 and 1640 by Sir William Pitt, is as modest as Blenheim is grand. It was originally of red brick, a long, low building with curved gables and small pediments. In the 18th century Lord Rivers stuccoed the brick, washing it over in a warm apricot colour, and entirely redecorated the interior. The Duke looked at several houses before deciding on Stratfield Saye in 1817, and though he originally planned to build a new house, 'Waterloo Palace', on a very grand scale near the present house, in the end nothing was built, as there simply was not enough money. He became fond of the house, though his friends considered it unworthy of him, and lived here quite modestly, only making minor additions, and building a conservatory at one end and a real tennis court at the other. There is a fine colonnaded hall containing Roman mosaics brought from Silchester; the other rooms, quite small, are all Georgian, though they are full of variety, as Lord Rivers' decorations spanned a period of some thirty years. The contents of the house are all Wellington's, and there is a fine collection of paintings, many brought back from Spain as spoils of war. The house has the feeling of a real home; all the Duke's personal possessions have been preserved, and the visitor can see such homely items as spectacles, handkerchiefs and carpet slippers lying about.

☎ Basingstoke (0256) 882882

11 m SE of Reading on A33 turn W for 2 m then S to Stratfield Saye House

SU 7061 (OS 175)

Open Easter to end Apr S and Su; May to end Sept daily exc F 1130-1700; reduced rates for parties

⊖ 🅿 WC 🅐 🍴 ♣ 🐕 ⅄ ◆ ⚹ ● (not in house)

The Vyne

Sherborne St John, Basingstoke, Hampshire

The Vyne is a Tudor house, but the name itself, meaning 'house of wine', is very old, and it is possible that there was once a Roman inn or villa on the site. The house, a long, low building of rose-red brick with diamond patterning, was built between 1500 and 1520 by the 1st Lord Sandys, Chamberlain to Henry VIII. It is a fine example of Tudor architecture, and many of its features, such as the tall, symmetrical windows, were innovatory. The long gallery (oak gallery) is one of the first to be found in an English country house, and there are outstanding examples of the work of foreign or foreign-influenced craftsmen, notably in the chapel, where the stained glass is Flemish, the choir stalls make use of Renaissance motifs and the tiles are probably from an Antwerp workshop. Sandys' descendants sold the house in about 1650, and the surprising Classical portico, also one of the first in the country, was added by the next owner, Chaloner Chute, who employed John Webb to improve the house. His descendant John Chute, a friend of Horace Walpole, made further improvements. The ground-floor rooms have Tudor woodwork, but the magnificent Palladian staircase was designed by John Chute, as was the tomb chamber. The rooms upstairs include the vast oak gallery; this room, untouched since Tudor times, has four rows of superb linenfold panelling from floor to ceiling, carved with the crests and initials of the King, Sandys himself and his relations and friends. The house is fully furnished, with some good Tudor portraits on the walls.

☎ Basingstoke (0256) 881337

3½ m N of Basingstoke on A340 turn E to Sherborne St John then N for 2 m

SU 6356 (OS 175)

Open Apr to late Oct daily exc M and F 1400-1800; Bank Hol M 1100-1800 (closed following T)

⊖ (1 mile) 🅿 WC ♿ 🚻 D (guide dogs only) ♣ ➽ ◆ ✂ 🏸 (by appt) ● NT

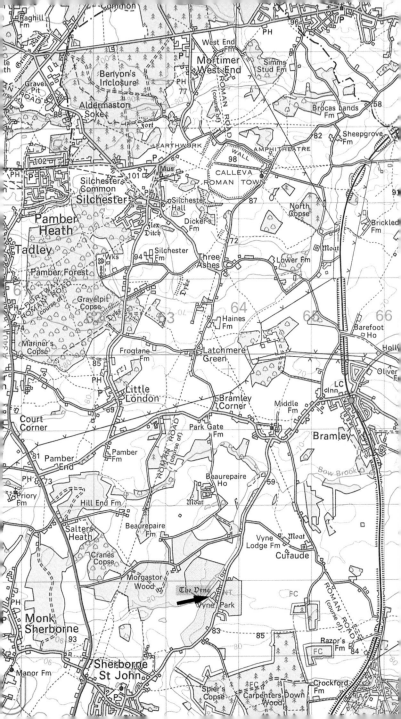

Osborne House

East Cowes, Isle of Wight

Queen Victoria, who had always loved the Isle of Wight, bought the estate in 1845, five years after her marriage. The house, which replaced a smaller one, was designed by Prince Albert himself with the advice of the architect Thomas Cubitt, and it is based on an Italian villa, with tall towers, a first-floor balcony and terrace gardens adorned with Renaissance-style statuary. When Prince Albert died in 1861 the Queen kept everything just as it had been in his lifetime, and her wishes are still respected. The public entrance to the house is through the extraordinary Durbar room, added in 1890 in honour of the Queen's Indian possessions and decorated with intricate Indian-style plasterwork. The three main ground-floor rooms are reached by the grand corridor, whose walls are lined with statues and pictures as well as cabinets containing gifts presented to the royal family. The billiard room has a table designed by Prince Albert, and in the drawing room is an unusual grand piano with ormolu mounts and porcelain plaques and allegorical statues of the royal children. The main staircase leads to the Queen's private suite. These rooms, in contrast to the grand ground-floor rooms, are full of personal possessions, including paintings by both the Queen and Prince Albert, and the Queen's sitting room, crowded with bric-à-brac, is typical of thousands of similar late-Victorian rooms. About half a mile from the house is the Swiss Cottage, a wooden chalet brought in sections from Switzerland and used by the royal children.

☎ Cowes (0983) 200022

SE of East Cowes, E of A3021

SZ 5194 (OS 196)

Open 28 Mar to end Oct M-S 1000-1700, Su 1100-1700

♿ 🅿 WC ♿ (limited access) 🚻 D (on lead, grounds only) ♠ ☛ ◆ ⁂ EH

Boughton Monchelsea Place

near Maidstone, Kent

A pleasant battlemented manor house on a fine site, with a southerly view of the edge of Romney Marshes. The battlements are in fact an early-19th-century fancy addition, but the house has a long history. The manor (Bolton) belonged to Earl Godwin in the 11th century, and after various Norman grants was held by the Montchensies (of which the name Monchelsea is a corruption) from near the end of the 12th century until 1287, and then by various Kentish families. In 1551 it was bought by Richard Rudstone, whose descendants held it continuously until 1888. Two wings of the originally four-sided house were taken down about 1740, and around 1790 the windows and the hall and red dining room were given the then fashionable 'Gothick' look. Other changes in the romantic taste, both to the house and the gardens and drive, followed in 1818-19, which gave them the appearance they have today. After 1888, the house lapsed into occasional lettings or was empty altogether for long periods, with the result that the kind of 'improvements' indulged in by many owners of other properties in the late 19th century passed Boughton Monchelsea by. In 1903 it came back into regular one-family ownership, but after the Second World War part was converted into flats. The fine staircase of 1685 replaced a more rustic Elizabethan one, the top flights of which can be visited in the upper floor. The Mortlake tapestries hanging in the house were originally hung in a room built for them at the top of the new staircase.

☎ Maidstone (0622) 43120

4½ m S of Maidstone on A229 turn E onto B2163 for 1 m then S

TQ 7649 (OS 188)

Open Good Fri to early Oct S, Su and Bank Hols, also W in Jul and Aug 1415-1800

♿ P WC ⊟ D ♣ ☞ ⊼ ◆ ⚶ ✗ (compulsory)
● (not in house)

Ightham Mote

Ivy Hatch, Sevenoaks, Kent

This 14th-century moated manor house, set in a quiet hollow and partly surrounded by woods, powerfully evokes the past. Its name comes, not from the moat, but from the 'moot' or local council, which was held here in medieval times. The buildings are a delightful mixture, with timber-framed upper storeys, red tile roofs and creamy-yellow stone below. The stone parts of the house are the earliest; the timber-framing and brick is probably 16th century. The house is entered by a bridge across the moat into a very attractive cobbled courtyard, and opposite is the oldest part of the house, a stone hall of about 1340. This still has the original doorway, but the oriel window was put in about 1500. The interior has a timber roof with corbels carved in the form of crouching figures, and the small two-light window is 14th century. The other outstanding early interior is the Tudor chapel, which has a barrel roof with painted decoration in the Tudor colours and fine woodwork on sanctuary, pulpit, screen and pews. The other rooms show a mixture of decorative styles from the Jacobean, through the 17th century to the Victorian, representing the long ownership of the Selby family (1598-1889), and there is some well preserved 18th-century Chinese wallpaper in the drawing room. Since their time the house has changed hands frequently, and was in danger of demolition when the present owner, Mr Robinson, fell in love with it, bought it, repaired and restored it, and now lives in it whenever possible.

☎ Plaxtol (0732) 810378

7 m N of Tonbridge on A227 turn W on road to Ivy Hatch and S to Ightham Mote

TQ 5853 (OS 188)

Open Apr to end Oct M, W-F, Su 1100-1700

🅿 WC ♿ (limited access) 🚌 (by appt) ♣ ✸
◆ ⚘ (for parties by appt) ● (not in house) NT

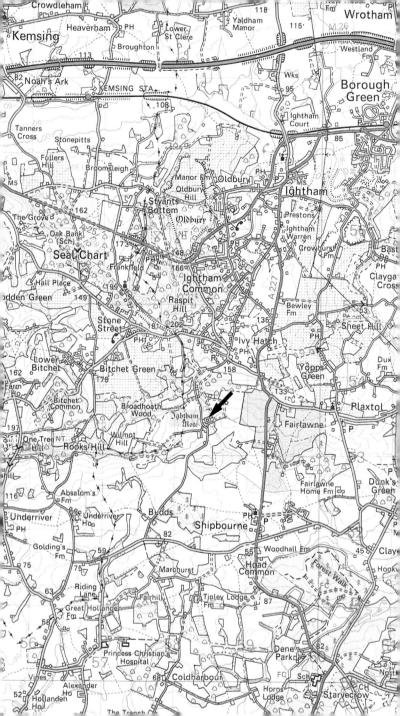

Knole

Sevenoaks, Kent

Set in a magnificent park studded with ancient trees, Knole is one of England's grandest houses. The vast house, more like a village, is popularly believed to echo the days of the year by having 365 rooms, set around seven separate courtyards. It has a complex building history, but owes its medieval appearance to Thomas Bouchier, Archbishop of Canterbury, who bought it in 1466 and transformed it into an archbishop's palace. It passed to the Crown in the 1530s when Henry VIII's covetous eye fell upon it, and was later given by Elizabeth I to Thomas Sackville, 1st Earl of Dorset, whose descendants still live in the private quarters. Henry had enlarged it by adding an outer courtyard (today's green court) and the long east front with its gatehouse, and Thomas made further alterations, notably creating a series of state rooms above the old palace. Little was done thereafter, as the Sackvilles deliberately cultivated the antique, and the huge grey stone house looks very much as it did in 1618. Knole is famous for the collection of furniture made by Thomas Sackville, and his splendid state rooms, with their lovely plaster-work, panelling and fireplaces, contain one of the finest collections of late 17th-century furniture anywhere in the country, including a set of silver furniture, a state bed with gold tissue hangings and the original 'Knole' sofa, dating from the time of James I. Many of the chairs still have their original upholstery, and the collection is in excellent condition.

☎ Sevenoaks (0732) 450608
1½ m S of centre of Sevenoaks on A225 turn E
TQ 5454 (OS 188)

Open Good Fri, Apr (or Easter if in Mar) to end Oct
W-S and Bank Hols 1100-1700, Su 1400-1700

⊖ (¼ mile) 🅿 WC ♿ (limited access) 🚻 D (on lead, grounds only) ♠ ⛺ ◆ ⚘ 🏹 ● NT

Pattyndenne

Goudhurst, Kent

A delightful old timber-framed house in the local style known as 'Wealden', in which parts of the upper storey and sides project as jetties, but the central part, having no jetties, gives the appearance of being recessed. This central part contained the hall, which was originally open to the roof. At Pattyndenne the whole upper storey rests on four moulded and chamfered corner posts, and the jetties project on all four sides. The plan of the house remains more or less unchanged: the central hall had a parlour on one side and a buttery on the other, with chambers above, and attics above these. When the house was built the only access from one end to the other was through the open hall. The house takes its Saxon name, which means 'a forest clearing by the stream', from the Pattyndenn family, who built it about 1480 as a house suitable not only for living in, but also for holding manor-court proceedings. In the 16th century it was sold to Sir Maurice Berkeley, son of the Lord Berkeley who was Standard Bearer to Henry VIII, Mary Tudor and Queen Elizabeth I. The house has four panes of glass showing the rose of Henry VIII and the pomegranite of Catherine of Aragon. The great fireplace in the hall was installed about 1580. The tiny kitchen wing was built about 1600, and there were no further alterations until 1890, when an extension was built at the back to take a new staircase. The present owners of the house, who bought it in 1972, have carried out considerable restoration work.

☎ Goudhurst (0580) 211361
13½ m E of Royal Tunbridge Wells on A262 turn S at Goudhurst onto B2079 for 1 m
TQ 7236 (OS 188)

Private parties only, by appt, throughout year

WC ♨ (by appt) ♣ ♨ (by appt) ✶ ✗ (by appt)

34

Penshurst Place

Penshurst, Kent

Among the few 14th-century domestic buildings to survive in England, the great hall at Penshurst is justly renowned. It was built early in the 1340s by Sir John Pulteney, a wool merchant and financier. The building consisted of the great hall with private apartments at one end and service quarters at the other. Built in local sandstone, with battlements, buttresses, arched windows and entrance porches, the hall is 62 foot long, with the apex of the roof about 60 foot above the floor. The roof timbers are a superb example of medieval woodwork. The great hall was surrounded by a wall interspersed with towers later in the 14th century. Only fragments of the walls remain, and without the original rectangular plan the layout of the later buildings now seem haphazard and confused. During the 1430s a new building, known as the Buckingham building, was added. It consisted of a crypt and one large room, but was divided up to form the forerunners of today's state rooms in the early 17th century. Edward VI granted Penshurst to Sir William Sidney in 1552, and his descendants have held the estate ever since. The long gallery was built by Sir Philip Sidney's brother Robert, who was also responsible for the division of the Buckingham building. Further extensive alterations were carried out by John Shelley, who inherited Penshurst through the female line and changed his name to Sidney in 1793. The apartments open to the public contain many fine furnishings and portraits, but it is the great hall that remains the outstanding feature.

☎ Penshurst (0892) 870307

2 m S of Tonbridge on A26 turn W on B2176 for 5 m

TQ 5244 (OS 188)

Open Apr to early Oct daily exc M (but inc Bank Hol M) 1300-1730 (grounds 1230-1800)

⊖ F WC ♿ 🛏 🍴 🍷 ⛽ 🪑 ◆ ⚬ ⚘ (by appt)
● (not in house) ⚘ ⚘

Squerryes Court

Westerham, Kent

A modest, symmetrical red-brick house of about 1680, looking out across a lake and small landscaped park. The outside, with its regular windows and matching pediments, is pleasant but plain, and the inside is equally so, reflecting the personalities of the squires who lived here for nearly 250 years. John Warde, son of a Lord Mayor of London, bought the house in 1731, and it now contains the Wardes' accumulation of family contents. There is a very good collection of Dutch and Flemish paintings made by John Warde II, which are hung in the three main downstairs rooms and the staircase hall, and many relics of James Wolfe (1727-59), captor of Quebec, came into the house through John's brother George, who was a friend of his. These are in one of the upstairs rooms. A room downstairs houses the regimental museum of the Kent and County of London Yeomanry, containing banners, military uniforms and so on. There is some good 18th-century furniture, some of which was made for John Warde's father, the Lord Mayor, before the family built Squerryes. The tapestry room contains some interesting decorative tapestries of about 1720 showing vases of flowers on marble tables, which were woven at the Soho factory by the well-known Joshua Morris. In the same room there is an original fireplace, one of several in the house.

☎ Westerham (0959) 62345/63118

7 m SW of Sevenoaks on A25 turn S in W outskirts of Westerham

TQ 4453 (OS 187)

Open Mar Su; Apr to Sept W, S, Su and Bank Hol M 1400-1800

♿ 🅿 WC 🚻 (by appt) D (grounds only) ♣ 🍽 (limited, by appt) ⊓ ♨ ✗ (by appt) ●

Chiswick House

Chiswick, London W4

This fine classical building with its clean, crisp lines was the creation of Lord Burlington, one of the most influential patrons of learning and the arts in the early 18th century, and a most accomplished architect. He was committed to a return to the ancient Roman architectural tradition as exemplified by the buildings of Andrea Palladio in Italy, and had little admiration for the current English styles, the ebullient Baroque of Wren and Vanbrugh. Chiswick House, which he designed in 1723, was based on Palladio's Villa Capra near Vicenza, and built in the grounds of his own house in Chiswick (now demolished). It was not intended as a house for living in, but as a 'temple of the arts', and the rooms were used to display paintings, sculpture and other works. The interiors were designed by William Kent, whose admiration for Palladian ideals matched Lord Burlington's own, though Kent proved to be rather less successful as an interior decorator than as an architect. Although the domed saloon and the gallery both have the grandeur and severity the building demands, in some of the other rooms the profusion of ornament seems to obscure the rigid lines and clean proportions. This is not helped by the fact that the original fabric which gave the red and blue velvet rooms their names has now been replaced with flock wallpapers. The gardens, which are as important as the house, were also designed by Kent, and show a break from the completely formal garden in which everything was laid out in straight lines.

☎ 01-995 0508

In Chiswick, just off A4 at Hogarth Roundabout

TQ 2077 (OS 176)

Open mid Mar to mid Oct daily 0930-1830; mid Oct to mid Mar W-Su 0930-1600. Closed 24, 25, 26 Dec and 1 Jan

⊖ 🅰 (limited) 🅿 D (grounds) ♣ ◆ ⚹ EH

Hampton Court Palace

Hampton, London

A dozen miles from Westminster, Thomas Wolsey's enormous red-brick mansion on the north bank of the Thames was built in the early 16th century when he was at the height of his power. The buildings were grouped round two main and several lesser courts, and were said to contain a thousand rooms. By the late 1520s Wolsey, falling out of favour, gave the palace and its contents to Henry VIII. The King enlarged the building, replacing Wolsey's great hall with the most impressive of all Tudor halls – over 100 foot in length, with a carved and moulded hammerbeam roof spanning 40 foot. Another main courtyard was added, with sets of apartments round it for the King and his Queens, and other quarters were extended and remodelled. The later Tudors and the Stuarts used the palace, but left few reminders of their presence. Cromwell retained it for his own use after the execution of Charles I. When William and Mary came to the throne in 1688 the building was outmoded, and Christopher Wren made major alterations. Work continued under William's successors until the death of George II in 1760, after which no sovereign occupied the palace again. Hampton Court has numerous delights: Wolsey's closet shows the luxury with which Wolsey furnished his country seat; in Wren's apartments for William III the King's staircase has walls and ceiling painted by Antonio Verrio; there are pictures, tapestries and furnishings, 15th-century cartoons by Andrea Mantegna, and gardens and grounds laid out at the time of Wren's alterations.

☎ (01) 977 1328

2 m W of Kingston upon Thames at junction of A308 and A309

TQ 1568 (OS 176)

Open Apr to end Sept M-S 0930-1800, Su 1100-1800; Oct to Mar M-S 0930-1700, Su 1400-1700

⊖ P WC ⬤ (limited access) ⊟ ♠ ⬤ ♫

◆ ⚘ ⚔ EH

Kensington Palace

Kensington, London W8

Kensington Palace, a rather unassuming and un-palatial building on the west side of Kensington Gardens, did not become a palace until William III decided that the air of Kensington might benefit his lungs, and moved here from Whitehall in 1689. It was then called Nottingham House, having been built for the Earl of Nottingham in 1661. Various additions were made to the old house by Sir Christopher Wren in the 1690s, and the process continued in 1720-21 during the reign of George I, when some of the interiors were lavishly redecorated by William Kent, who also laid out the gardens. The orangery and the King's gallery, built in 1695 and 1704 respectively, may have been designed by Nicholas Hawksmoor, and the orangery has a very fine interior with carved panels by Grinling Gibbons. The state apartments, which have been open to the public since 1899, have been returned to their 18th-century appearance, with suitable paintings and pieces of furniture from the Royal Collection, and the painted decorations by William Kent have been restored. It is rather a surprise amid the 18th-century elegance to come to a suite of rooms entirely in the Victorian style, but Queen Victoria was born in Kensington Palace and was very fond of it. The last room to be seen by visitors is the most splendid of the 1720 additions, the cupola room. This was intended to be the main state room, and is decorated with pilasters and gilded statues of Roman gods and emperors.

☎ (01) 937 9561

In Kensington on Palace Avenue off Kensington High Street

TQ 2580 (OS 176)

Open throughout year M-S 0900-1700, Su 1300-1700. Closed 24,25,26 Dec and 1 Jan

⊖ 🅿 WC 🅰 🍴 ♿ 🄑 ◆ ☼ ● (not in house) EH

Kenwood House

Hampstead Lane, London

This elegant mansion on the northern edge of Hampstead Heath was built around 1700, but owes its appearance to Robert Adam who remodelled it in 1764 for the Earl of Mansfield. He designed the grand entrance portico and pediment on the north side, and redesigned the entire south front overlooking the parkland, building a library on the east side to balance the existing orangery on the west. In 1793 the 2nd Earl caused the main Hampstead-to-Highgate road to be moved to its present position (it originally ran closer to the house) so that the house could stand surrounded by its own parkland, and the entrance (north) front is now approached by two winding drives through woods. The two wings, one each side of the entrance portico, were added by George Saunders in the 1790s. The house was bought by Lord Iveagh in 1925, and left to the nation, together with his superb collection of paintings, in 1927. The house is mainly used as a 'display case' for the Iveagh Bequest paintings, and although most of the main rooms were decorated by Adam and still contain some very fine pieces of furniture, the rooms are arranged like a museum, not an occupied country house. The one exception is the library, which is more or less as Adam left it and is one of his finest interiors. The Iveagh Bequest paintings, which include works by Rembrandt and Vermeer as well as outstanding portraits by Gainsborough and Reynolds, are on permanent display in the downstairs rooms, but the upstairs rooms are sometimes used for temporary exhibitions.

☎ (01) 348 1286

On Hampstead Heath off B519 Hampstead Lane

TQ 2787 (OS 176)

Open daily, Apr to end Sept 1000-1900; Mar and Oct 1000-1700; Nov to Feb 1000-1600

♿ P WC 🦽 (limited access) 🚻 (by appt)
D (grounds only) ♣ ♟ ☂ ★ ◆ ✳ ● ☂

40

Lancaster House

Near St James's, London SW1

Lancaster House is an imposing Georgian building with a characteristic buff-coloured exterior of Bath stone. It stands in the heart of royal London, across the road from St James's Palace and just off the Mall leading to Buckingham Palace. It was first designed for King George IV's brother Frederick, Duke of York, by Sir Robert Smirke; but Smirke was superseded by Benjamin Wyatt, who produced new designs and began building 'York House' in 1825. The Duke died in 1827, and the still unfinished building was bought by one of his creditors, the Marquess of Stafford (later Duke of Sutherland). As Stafford House it was long a notable centre of politics, fashion and art, so lavishly furnished that when Queen Victoria arrived she would tell the Duchess: 'I have come from my house to your palace'. In 1912 Viscount Leverhulme bought the house and renamed it Lancaster House in honour of his native country – with the result that, in the course of its history, it has represented both sides in the Wars of the Roses! Leverhulme presented it to the nation, and it is now used as a conference centre. The interiors, mainly by Wyatt and Sir Charles Barry, are still sumptuous, using deeprich colours in an overall setting of white and gold. Outstanding features include the central staircase hall, rising the full height of the building to a coved ceiling supported by caryatids and carrying a huge lantern; Wyatt's grand staircase; and the great gallery on the first floor, over 120 foot long and filling one side of the house.

☎ (01) 212 4784

In Stable Yard, St James's, SW1

TQ 2979 (OS 176)

Open Easter to mid Dec S, Su and Bank Hols 1400-1800; closed for Government functions

♿ ⬛ (ground floor only) ⊟ (by appt) ★ ⚹ EH

41

Southside House

Wimbledon Common, London SW19

This is a pleasant, sturdy house with a long Classical façade crowned by two pediments, square dormer windows and a clock tower. It was built for Robert Pennington, who had shared Charles II's exile in Holland and evidently acquired a taste for Dutch architecture. When the plague carried off his small son he left London with his wife and daughter, and retired to Holme Farm at Wimbledon, which was then still a village several miles from the capital. In 1687 he called in Dutch architects to build Southside House, incorporating the farm building in the structure; this accounts for the asymmetrical placing of the front door and clock tower. Two very large niches, one on each side of the front door, were filled with statues of Plenty and Spring, which are still in place; the faces are said to be likenesses of Pennington's wife and daughter. Southside House remained a family home over the centuries with only one set of major building alterations (in 1776); and Pennington's descendants lived there until after the Second World War. The interiors include a good deal of 17th-century furniture and a range of memorabilia connected with the Penningtons. The 'musik room' was prepared for the entertainment of Frederick, Prince of Wales, who stayed at Southside House in 1750, and later visitors included one of history's most famous *ménages à trois*, Sir William and Lady Hamilton and Lord Nelson. Southside House was damaged during the Second World War, after which it was restored to its late 17th-century condition.

☎ (01) 946 7643

On S side of Wimbledon Common near Crooked Billet inn

TQ 2370 (OS 176)

Open Oct to end Mar (exc Christmas and Easter) T, Th, F 1400-1700; by written appt at other times

♿ WC 🚻 ♨ ✗ (compulsory) ● (permission required)

Marble Hill House

Richmond Road, Twickenham, Middlesex

Marble Hill, a small and beautifully proportioned Palladian villa, was built for Henrietta Howard, Countess of Suffolk – mistress of George II – as a country home. The house was completed in 1729, and as the Palladian style grew more popular it became one of the 'standard models' for the 18th-century villa. The architects appear to have been a partnership consisting of Colen Campbell (who produced the initial designs), Roger Morris and Lord Herbert, connoisseur, amateur architect and contemporary of that great arbiter of taste, Lord Burlington. The house still looks very much as it did when first built, despite many years of neglect in the late 19th century; the interiors of the charming small rooms have been restored and refurnished in the 18th-century style. The hall contains four Ionic columns dividing the room. The most spectacular room is the great room which occupies the centre of the river front, and has lovely carved and gilded ornament and a coved ceiling. It was probably modelled on the cube room at Wilton, built by Inigo Jones. The house is administered by the Greater London Council and some of the 18th-century furniture and paintings have been acquired recently, but in spite of their efforts the house has a rather 'un-lived-in' feeling. The large riverside park which once belonged to the house, and which was laid out by Charles Bridgeman with the help of Alexander Pope, is now a public park.

☎ (01) 892 5115

1 m E of Twickenham, S of A305 Richmond Road

TQ 1773 (OS 176)

Open daily exc F; Feb to end Oct 1000-1700; Nov to end Jan 1000-1600. Closed 24, 25 Dec

⊖ P WC ⴲ (limited access) ⛺ (by appt) ♣
🐾 (by appt) ⛱ ★ ◆ 🍴 ⚲ ⚐ ● EH

Osterley Park House

Osterley, Isleworth, Middlesex

Like Robert Adam's other great masterpieces in the London area, Kenwood and Syon, Osterley is a remodelling of an existing house, in this case an Elizabethan one built in the 1570s for Sir Thomas Gresham. In the mid-18th-century Sir Francis Child, then the owner, had already begun rebuilding (his architect was probably Sir William Chambers), and the entire house was refaced, a long gallery built right across the west front, and various other alterations made. Adam took over in 1761, and his first work was the great pedimented portico which filled the open side of the original courtyard. He then began to redecorate and refurnish seven of the nine ground-floor rooms, and these state rooms, completed in 1780, remain almost exactly as he left them. The architecture of the interiors is dignified and perfect, but the rooms were designed for show rather than comfort, and some of them seem rather austere. As far as possible all the furniture, made to Adam's own designs, has been placed as he intended, standing against the walls as was the fashion in the 18th century, and complemented by paintings of the period. At the end of the tour, in the last room and the passage leading from it, there is a display of Adam's drawings for the furniture and decorations. The large park, with its lakes and ancient yew trees, contains a semi-circular garden house by Adam and a Greek temple which was probably designed by Chambers. The rustic stable block, now a tearoom, is believed to be the remains of the original Tudor house.

☎ (01) 560 3918

In Osterley S of M4 on Great West Road (A4), turn N onto Syon Lane and Osterley Lane

TQ 1478 (OS 176)

Open daily exc M (but inc Bank Hol M) 1100-1700

⊖ P WC ♿ (limited access) 🚻 D (guide dogs) ♣ 🍴 (Apr to Oct) 🏕 ◆ ✿ 🍴 (by appt) NT

Syon House

Syon Park, Brentford, Middlesex

This great house, set in its tree-dotted park, is surprising to find in London's suburbia. Syon takes its name from a monastery founded by Henry V in 1415. After the Dissolution the lands were given to the Duke of Somerset, who built himself a large house, retaining some of the monastic buildings. Basically this is the house we see today, although it was refaced with Bath stone and sash windows put in, which gives it an 18th-century look. Its castellated exterior is not very beautiful but the interior more than makes up for it. This is the house that made Robert Adam's name, and although he was only twenty-nine, the series of rooms he designed here are among the most brilliant of his career. The visitor sees the rooms in the order that Adam intended, starting in the stately Roman-style hall, where the plasterwork was done by Joseph Rose who also made the statuary to Adam's designs. The ante-room, with its gilding and its vivid colouring, is straight out of the Roman Empire at its most lavish. The dining room, the first room Adam completed, is mainly ivory and gold, while the red drawing room takes its name from the crimson Spitalfields silk on the walls. A fine Moorfields carpet designed by Adam fills the room, and the ceiling is an elaborate pattern of coloured octagons and diamonds intersected by bands of gilding. The final room in the series, the long gallery, was planned by Adam as essentially a room for the ladies. The whole room is in pale greens and gold-spangled pinks, and sparkles with delicate ornament.

☎ (01) 560 0881

In Brentford S of London Road A315 in Syon Park

TQ 1776 (OS 176)

Open Good Fri to late Sept Su-Th 1200-1700; grounds open 1000-1800 or dusk if earlier

⊖ Ⓟ WC 🚻 ♥ 🍴 ◆ ♨ ⚹ ⚔ ● (not in house)
Butterfly house; aviary; aquarium etc

Detillens

Limpsfield, Surrey

The house, which takes its name from an 18th-century owner, James Detillen, was probably built about 1450. Its early Georgian front, added about 1725 in accordance with current taste, conceals a 15th-century timber-framed house. The earliest known occupier was Richard Kinge, a lawyer, who lived here around 1600, and at some stage the house was evidently owned by the local miller, since a quantity of flour has been found between floorboards and rafters. Originally the centre of the house would have been the large hall, open to the roof, but early in the 16th century this was divided both vertically and horizontally to form the front hall and dining room downstairs and bedrooms upstairs. The roof of the old hall had a central truss resting on a huge tie-beam, and this can still be seen in the main bedroom. The front hall now displays an excellent collection of military and sporting guns. The study, which was once the solar, or main sitting room of the house, has a fine Elizabethan overmantel bearing the figures of three caryatids, and the names of two previous occupants of the house are carved into the wood. The staircase is Jacobean, and the Tudor morning room has a fireback bearing the arms of Elizabeth I. This room also houses a selection from a unique collection of orders and decorations, both British and foreign, which has been amassed over the years. The present owners, who bought the house in 1968, have collected all the contents as well as restoring the house and the topiary garden, which is also open to the public.

☎ Oxted (088 33) 3342

9½ m SW of Sevenoaks on A25, turn NW on B269 for ¼ m

TQ 4052 (OS 187)

Open May and June S and Bank Hols; July to end Sept W, S and Bank Hols 1400-1700; parties at other times by appt

⊖ 🅿 🍴 ♿ ✕ 🚶 (compulsory) ● (not in house)

Ham House

Ham, Richmond, Surrey

Originally an H-shaped Jacobean house built in 1610 for Sir Thomas
Vavasour, Ham House was completely redecorated in the 1670s, and is now
the finest existing example of the interior decoration of the Restoration
period. The 1st Earl of Dysart made some alterations in the 1630s,
redecorating the older rooms and inserting the staircase, but it was his
daughter Elizabeth and her husband, the Duke of Lauderdale (1616-82), who
made the house we see today. They enlarged it by adding a new range of
rooms, thus filling in the south side of the H, but their work on the exterior
was minimal compared to the time, money and skills lavished on the interior,
particularly the new rooms. The ground floor contains the private apartments
including bedrooms, an arrangement we find rather odd today; and most of
these are panelled or hung with leather. The great staircase (*c.* 1637), with
grained and once-gilded panels and an elaborately carved balustrade, leads
to the first-floor state rooms which, although relatively small in scale, are
furnished with incredible splendour, with painted ceilings, parquetry floors,
marble chimneypieces and tapestry, damask or velvet wallhangings. Much
of the furniture has no parallel in the country and was specially made for the
house, and many of the pieces are mentioned in an inventory of 1679. In
several of the rooms the paintwork and fabric have been renewed with the
utmost care, and the bright colours (which some people dislike) are those
which would have been used originally.

☎ (01) 940 1950

W of Richmond Park at Petersham off A307
on Ham Street

TQ 1773 (OS 176)

Open daily 1100-1700 exc M but inc Bank Hol M.
Closed Good Fri, 25, Dec, 1 Jan and early May Bank
Hol M

♿ 🅿 WC ♿ 🚻 ♣ 💷 ◆ ♨ 🍴 (W only) NT

Kew Palace

Kew Gardens, near Richmond, Surrey

The building now known as Kew Palace, formerly known as the Dutch House, was built in 1631 by a leading London merchant called Samuel Fortrey. He had Dutch connections, and the tall red-brick house with its curly gables and brick ornament is in the Netherlandish style then popular in England; hence its name. The house is only called 'palace' because by the end of the 18th century it had become part of the royal estates at Kew and Richmond, and a new palace, of which nothing now remains, was built adjoining it. The principal royal residence was then a house called the White Lodge, which was pulled down in 1802 to make room for the new palace (which was never completed). King George III and Queen Charlotte stayed temporarily at the Dutch House, and Queen Charlotte remained there until her death in 1818. All the ground- and first-floor rooms are open to the public, and have been restored to more or less their Georgian appearance, though there is still some of the original 17th-century plasterwork in the Queen's boudoir. One of the rooms has an attractive collection of royal ephemera and playthings. The small garden between the house and the river was first planted in 1696 by Sir Henry and Lady Capel, then owners of the house, and has been restored to give the appearance of a 17th-century garden. The botanical gardens themselves, given to the nation in 1841, were landscaped by Capability Brown for George III and his mother Princess Augusta.

☎ (01) 940 3321

In Kew Gardens, W off Kew Road to Kew Green

TQ 1877 (OS 176)

Open Apr to Sept daily 1100-1730

⊖ WC ♿ (ground floor only) 🚻 D (guide dogs only) ♦ ☞ ⌗ ◆ ✄ ● (not in house) EH

Loseley House

Loseley Park, Guildford, Surrey

A gabled Elizabethan house, built of greenish-grey ragstone from the ruins of nearby Waverley Abbey, with window surrounds of hard white chalk, Loseley House was built between 1562 and 1568 for Sir William More, a kinsman of Sir Thomas More, and a man respected by leading statesmen and trusted by Elizabeth I, who stayed here three times. Sir William's descendants, who married into the Molyneux family, still live here, and now run a well-known dairy farm on the estate. There have been few alterations to the exterior of the house, though the main doorway appears to be a Queen Anne addition. The great hall, a high room with a fine oriel window, heraldic glass and a beamed ceiling, has some good portraits including a large family group painted by Van Somer in 1739. There is also a remarkable series of carved, inlaid and painted panels in the Italian style, which are believed to have come from Henry VIII's palace at Nonsuch (destroyed after the Restoration). The library has some good 16th-century panelling, but most is 19th century, while the drawing room has a plaster ceiling and a striking chimneypiece carved out of chalk. The staircase is 17th century, and the bedrooms upstairs have 16th-century ceilings. The house contains a good collection of furniture as well as some tapestry and needlework, one of the most unusual pieces being a 16th-century German marquetry cabinet with a design showing a fallen city.

☎ Guildford (0483) 571881

4 m S of Guildford on A3100, turn W onto B3000 for ¾ m, then N to Littleton

SU 9747 (OS 186)

Open June to end Sept W-S 1400-1700

⊖ (1½ m walk) P WC 🚻
D (guide dogs only) ♟ 🍴 🗦 ◆ ⚹
𝒦 (compulsory) ● (not in house)

Polesden Lacey

near Dorking, Surrey

This delightful house, set in the pleasant countryside of the North Downs, was the home of the well-known hostess Mrs Greville, and has seen many distinguished visitors including Edward VII and the Duke and Duchess of York (later George VI and Queen Elizabeth). Originally a Georgian villa, the present house was begun in 1824 to designs of Thomas Cubitt on the site of an earlier house which belonged to the playwright Richard Brinsley Sheridan. Although it was greatly enlarged early in this century it retains something of its Regency flavour, and the south front with its colonnaded portico has been little altered except for an extension to the east. The enlargements were made in 1906 by Ambrose Poynter for the Hon. Ronald Greville and his wife, and the east front is nearly all his work. Mrs Greville was an art collector as well as hostess, and the opulent rooms contain many treasures. The dining room, with rich brocade hangings, contains her collection of English portraits, but the major part of the art collection, which includes important Flemish and early Italian works and Dutch landscapes, hangs in the corridor around the central courtyard. This leads to the airy library, which has been little changed since Mrs Greville's day, and then to the sumptuous drawing room, where the walls are covered in carved and gilt panelling. This room contains some particularly fine furniture and oriental porcelain. The attractive gardens were planted by Mrs Greville, though the long terrace was laid out by Sheridan in the late 18th century.

☎ Bookham (0372) 58203/52048

2 m N of Dorking on A24, turn W to Westhumble

TQ 1352 (OS 187)

Open Mar and Nov S, Su 1400-1700; Apr to end Oct daily exc M and T 1400-1800 (open Bank Hol M)

🅿 WC ♿ 🚻 (by appt) D (on lead; not in formal gardens) ♣ ♥ 🗚 ◆ ⚜ ● (no video) NT

Whitehall

Malden Road, Cheam, Surrey

Built about 1500, Whitehall was probably a yeoman farmer's house attached to East Cheam Manor which was, in turn, owned by the See of Canterbury. In 1539, when the nearby Nonsuch Palace was being built, Henry VIII bought the manor from Archbishop Cranmer, and Whitehall was reputed to have been used to house the Maids of Honour when the King was at Nonsuch. In 1654 Queen Mary granted the manor to Arthur Browne, Viscount Montague and in 1575 it was sold to the Earl of Arundel. From 1645 to 1719 it was the home of Cheam School. Sutton and Cheam Borough Council acquired the property in 1963 and have been responsible for the very careful restoration work. The white-painted weatherboard style of Whitehall is not uncommon in this part of the country but the timber-framed continuous-jettied house has some unusual features, notably a projecting upper storey and a two-storey porch with an attractive depressed Tudor arch. The original house was of two storeys, with three rooms – a hall, stillroom and kitchen-parlour – on the ground floor and a sleeping room and storeroom upstairs. Extensions were added later in the 16th century and also in the 17th and 19th centuries; the weatherboarding itself is a late 18th-century addition.

☎ (01) 643 1236

In centre of Cheam off Malden road

TQ 2463 (OS 176)

Open throughout year W-Th, Su and Bank Hol M 1400-1730, S 1000-1730; also T pm Apr to end Sept

♿ 🅿 WC ♿ (limited access) 🚃 (by appt)
♦ 🍴 ◆ ✾ ✗ (by appt)

Firle Place

near Lewes, East Sussex

The house lies in its fine parkland beneath the slopes of Firle Beacon. It is the home of the Gage family, and was built for Sir John Gage in the mid-16th century, remaining more or less unchanged until the 18th century. Extensive additions and alterations were made between 1713 and 1754, and although a gable on the south side still survives from the Tudor building and the plan, with its two courtyards, is probably the original one, the old house was otherwise almost entirely swallowed by the new one. The approach is through one of the courtyards, and the Classical front door opens straight into what was once the Tudor great hall. Little that is Tudor now remains even here, however: in the 18th century the room was divided into two, an elegant entrance hall and a superb Palladian staircase hall. One of the finest rooms is the drawing room, also Palladian, with screens of Ionic columns and full-length family portraits set in white and gold panelling. There is also some fine furniture, but the best furniture and paintings are in the upper rooms. Here there is a magnificent collection of mainly French 18th-century furniture, including a roll-top desk in the manner of Riesener, a famous collection of Sèvres porcelain, and several important paintings including Fra Bartolomeo's *The Holy Family and the Infant St John*. The long gallery, built about 1713, and with a fine spectacular view of the South Downs, contains excellent portraits and paintings of the English school, as well as a major work by David Teniers, *The Wine Harvest*.

☎ Glynde (079 159) 335

6 m E of Lewes on A27 turn S to West Firle

TQ 4707 (OS 198)

Open Jun to Sept W, Th, Su and Easter, May and Aug Bank Hols Su, M 1415-1700

⊖ (1½ m) P WC ☒ (limited access)
🍴 (by appt) ♠ ☕ 🚻 ◆ ⚘ ✗ ●

Michelham Priory

Upper Dicker, Hailsham, East Sussex

As its name implies, this Tudor house was built on to an old priory. This had been founded in 1229; the buildings were gradually put up over the next hundred years, and the great moat, which encircles today's house, was dug slightly before 1400. The gatehouse was built at the same time, but the bridge leading to it is 16th century. After the Dissolution in 1536 the buildings were destroyed, but later in the century they were repaired and incorporated into a Tudor house which became the centre of a working farm, owned by the Pelham family. The priory buildings were arranged round a cloister, and the house we see today, which was converted into a 'gentleman's residence' early in this century, consists of the south-west corner of the cloister with the Tudor wing at the side. Michelham Priory is now a property of the Sussex Archaeological Society, who have made an effort to illustrate the way of life of the Augustinian monks. Near the entrance is a model of the priory as it probably appeared on completion, and all the priory rooms contain life-size models of monks. The other rooms, arranged to look as though they are still in use, contain 17th-century furniture, tapestries and so on. The 16th-century kitchen still has its great open hearth and some of the original oak beams, while the prior's guest room has a fireplace dating from about 1320. The oak panelling was installed in the 19th century. Two of the rooms are used as a museum, one concerned with local archaeology and the other containing a collection of musical instruments.

☎ Hailsham (0323) 844224

9 m N of Eastbourne on A22 turn W to Upper Dicker for 2½ m

TQ 5609 (OS 199)

Open 25 Mar to 31 Oct daily 1100-1730

🅿 WC ♿ D (on lead, grounds only) 🍴 ♣ ☕
⋒ ◆ ⚲ 🏹 ● (no tripods) ✶

Forge and working water mill; August Arts Festival

Royal Pavilion

Brighton, East Sussex

At the end of the London road into Brighton, the Royal Pavilion must have startled travellers when its highly original domes and minarets first rose skywards. The Pavilion had its origins in a modest farmhouse near the sea front, which Henry Holland enlarged for George, Prince of Wales, son of George III, in the 1780s. Much of Holland's Neo-Classical structure is contained within the present building. The interior was redecorated, at huge expense, in the Chinese style between 1801 and 1803, and during this time a stable block was constructed in the grounds in Indian style. The Prince was delighted with it, and the Indian style was adopted again in the construction of the present Pavilion, on which work began in 1815 under John Nash. Completed in 1822, the Pavilion was a unique building, a glorious fantasy based on eastern themes. Thanks largely to the Prince's patronage, Brighton had become an increasingly popular resort. Queen Victoria liked neither the crowded town nor the Pavilion, which she found odd. In 1845 she purchased Osborne, on the Isle of Wight, for her seaside residence. The Pavilion was damaged when furniture and fittings were removed to royal storerooms, and was only a shell when the town bought it in 1850. It remained shabby until after the Second World War, when the sophisticated restoration work was begun that has resulted in a wonderful authentic return to the inventive, exuberant Pavilion of 1822.

☎ Brighton (0273) 603005

In centre of Brighton near the Palace Pier

TQ 3104 (OS 198)

Open spring to end Sept daily 1000-1800, Oct to spring 1000-1700

♿ WC 🅿 🚻♿ 🍴 ♦ 🎁 ✂ (available in foreign languages by appt) ● (no tripods)

Sheffield Park

Uckfield, East Sussex

The house, an early example of the 18th-century 'Gothick' style, is largely the work of the architect James Wyatt. The original house acquired by Lord Sheffield was 16th century, and in 1775 Wyatt was engaged to rebuild it. By 1778 his work was complete, and the east front, facing the lakes, survives virtually unchanged. The west front was given battlements and turrets towards the end of the century, and further additions, notably the delightful orangery, were made by a subsequent owner, Arthur Gilstrap Soames, in 1912. Wyatt's marvellous, huge Gothick window on the east front, the focal point of the house, was put there for artistic effect alone, as on the inside it is blocked by lath and plaster. Unfortunately the house and garden now have different owners – the garden belongs to the National Trust – so in order to walk round the outside of the house as well as the inside one has to pay twice. The present owners of the house, Mr and Mrs Radford, bought it in 1972 totally unfurnished and partially undecorated, and they have been carefully restoring it ever since. Two of its best features are the lovely vaulted staircase hall designed by Wyatt and the first-floor bedroom known as the 'tyger room', which has a ceiling painting showing tigers, lions and leopards done by Charles Catton about 1778. The large triple bay window gives a superb view of lawn and lakes, and there is a fine fireplace. The spacious library where Edward Gibbon is said to have written some of *The Decline and Fall of the Roman Empire* has its original bookcases and marble fireplace.

☎ Dane Hill (0825) 790655

11 m N of Lewes on A275 just before Sheffield Green

TQ 4124 (OS 198)

Open Apr to mid Nov T-S 1100-1800; Su and Bank Hol M 1400-1800 (sunset if earlier); closed Good Fri

P WC 🅿 🖥 D (guide dogs only) ♠ ☛ ⊓ ◆
🌿 🎋 (twice a year, NT members only) NT

Goodwood House

near Chichester, West Sussex

The 18th-century house was originally a hunting lodge, which was bought in 1677 by the 1st Duke of Richmond and greatly enlarged to the designs of James Wyatt by the 3rd Duke. The elegant stable block was designed by Sir William Chambers and built in the 1750s, and about twenty years later the Duke began to plan a lavish new house, to be a setting for entertainments, concerts and so on. Wyatt supplied designs, but of these only the tapestry room and kennels (now a golf club) were built, and it was not until about 1800 that the Duke and Wyatt thought again about rebuilding the house. This time the plan was for a great octagon, arranged round a central courtyard containing the older house; but this extremely bold architectural plan never came to fruition, and only three sides were completed. These are built of flint, and are three storeys high with round towers at the corners. The rooms themselves are rather austere, and all except the tapestry room were redecorated in 1970 to form a neutral background to the very fine collection of French furniture, Sèvres porcelain and clocks brought back by the 3rd Duke, who was an ambassador at the court of Louis XV. Wyatt's tapestry room is a fine example of his decorative style, and is hung with superb 18th-century Gobelin tapestries of Don Quixote. The unusual marble chimneypiece was made by John Bacon. The house contains many fine paintings, including several portraits by Van Dyck and Lely, two Canaletto views of London and equestrian paintings by Stubbs and Wootton.

☎ Chichester (0243) 774107

3 m NE of Chichester on A27, turn N to Waterbeach

SU 8808 (OS 197)

Open Easter Su and M and May to mid Oct Su and M, Aug Su–Th 1400–1700

 (not in house)

Parham Park

Pulborough, West Sussex

This simple grey stone Elizabethan house began to be built in 1577 by Sir Thomas Palmer. Palmer, who went to sea with Drake, evidently preferred sea to land, and in 1610 Parham was sold to the Bysshopp family, branches of which retained it for eleven generations. In 1922 it was bought by the Hon. Clive Pearson, whose daughter and son-in-law now own it. All Parham's owners, including the present ones, have cherished the house, and alterations and restorations have always been carried out with tact and taste. The buildings to the north of the house are 18th century; the range nearest to it was much enlarged in the 1770s, and about 1800 the main entrance was moved from the south porch to the north side and an entrance hall built. These alterations had almost no effect on the interiors, which are nearly all early, with carved panelling and Tudor and Jacobean furniture. The splendid great hall, lit by tall windows, has its original stone fireplace and Renaissance carved screen, and the decorated ceiling, though a 19th-century copy of the original, gives a fine 16th-century effect. The very long gallery (160 feet in length) has its original floor and carved oak wainscot, and the delightful ceiling is a modern interpretation of an Elizabethan theme. The great parlour and great chamber above it are examples of superb restoration work: Edmund Burton did the plasterwork in 1935, using the old method of modelling with the fingers. Parham's paintings include a series of historical portraits and a famous painting of Queen Elizabeth, possibly by Zucchero.

☎ Storrington (090 66) 2021

3 m S of Pulborough on A283

TQ 0614 (OS 197)

House open Easter Su to 1st Su in Oct, W, Th, Su and Bank Hols 1400-1800, gardens open 1300-1800

P WC ♿ (limited access) 🚻 (by appt) D ♣ ☛
🚗 ◆ ✕ 🏇 (by appt M pm, W and Th am) ●

Petworth House

Petworth, West Sussex

Petworth is famous for its park, the masterpiece of Capability Brown, and its superb collection of paintings, among which are nineteen by J. W. M. Turner, who spent much time here in the 1830s. But although it may be primarily an art gallery, it is also a very impressive country house. The house originally belonged to the Percy family, but almost nothing can now be seen of the medieval building except a 13th-century chapel. In 1682 Charles Seymour, 6th Duke of Somerset, married the Percy heiress, and the present house was built by him between 1686 and 1696. The long main front is very French in appearance, but the architect is not recorded. The visitor enters through the east front, which was partly rebuilt by Anthony Salvin in 1870. The ground-floor rooms are all very opulent, though the only room to survive from the 6th Duke's time is the marble hall. The richly carved decoration is full of French and Dutch features, and it is thought that a Huguenot craftsman and designer, Daniel Marot, may have made the designs. The square dining room, hung with Van Dycks, is among the rooms created by Lord Egremont early in the 19th century. The beauty room, so called because it is devoted to the ladies of Queen Anne's court, has portraits by Kneller and Dahl, while the staircase hall is decorated with murals by Laguerre. The carved room contains a riot of woodcarving by Grinling Gibbons. There is a collection of ancient sculpture made by Turner's patron Lord Egremont, together with his collection of contemporary paintings.

☎ Petworth (0798) 42207

8 m E of Midhurst on A272

SU 9721 (OS 197)

Open Apr to end Oct daily exc M and F (but inc Bank Hol M) 1400-1800

🅿 WC ♿ ➍ (by appt) ♠ ♣ ◆ ⚘ NT

Uppark

South Harting, near Petersfield, West Sussex

This attractive house, built on the top of the South Downs, has been nicknamed the 'Sleeping Beauty House' because for most of the 19th century it was occupied by two old ladies who refused to allow any changes. There are many human dramas and stories attached to Uppark. Sarah Wells, the mother of H. G. Wells, was housekeeper here from 1880 to 1893. The house was built in the 1690s for Lord Tankerville, possibly by William Talman, and is typical of houses of this date – built of brick, and three storeys high with a central pediment, the upper storey lit by dormer windows in the roof. It was bought in 1747 by Sir Matthew Fetherstonehaugh, and he and his wife altered and completely redecorated it, leaving only the hall, staircase and dining room as they had been. The exterior was not altered at this time, but the Doric colonnade on the north front was built by Sir Humphry Repton in 1810. The furniture is mainly 18th-century English, and there are Dutch and Italian paintings mostly collected by either Sir Matthew or his son Sir Harry, who turned the house into a favourite haunt of the rich and dissolute society surrounding the Prince Regent. In 1825 Sir Harry, then aged over seventy, married his dairymaid, Mary Ann Bullock – the two ladies who were to preserve the house so well were her sister and former governess. Upstairs the house still has the comfortable feeling of an old ladies' home, while downstairs the housekeeper's room has been arranged just as it was in Sarah Wells' time, with a photograph of her encased in black satin.

☎ Harting (073 085) 317/458

7 m SE of Petersfield on B2146

SU 7717 (OS 197)

Open Apr to end Sept W, Th, Su and Bank Hol M
1400-1800

⊖ (1½ m) P WC ♿ (limited access)
🚌 (by appt) ♣ ⬛ ◆ ☀ NT

Berkshire

Basildon Park, near Pangbourne, Berkshire (tel Pangbourne [073 57] 3040). 9 m NW of Reading on A329, turn W. SU 6178 (OS 175). Attractive Palladian house built in the 1770s by John Carr in a park overlooking the River Thames. The staircase is an important feature of the house. Much of the decoration is original, and the rooms are furnished in 18th-century style. There is an unusual octagon room, completed in the 1840s. Open Apr to Oct W-S 1400-1800, Su and Bank Hol M 1200-1800; closed W following Bank Hol. ⊖ 🅿 (400 yds uphill to house) WC 🅰 (limited, by appt) 🚻 (by appt, not Su, Bank Hol M) D (on lead, grounds only) ♠ 🍴 🍴 🏕 ◆ ✹ ✗ (by appt) ⬤ (not in house) NT

Swallowfield Park, Swallowfield, Berkshire (tel Reading [0734] 883815). 5 m SE of Reading on A33, turn SW onto B3349 for 4 m, then turn E. SU 7365 (OS 175). House built c. 1690 for the 2nd Earl of Clarendon by William Talman, architect of Chatsworth, and reworked in the 1820s. Features include the original Baroque doorway, now moved from its original position, and a fine four-acre walled garden. Open W and Th pm, May to Sept. ⊖ 🅿 WC ♠ ✗

Hampshire

Alresford House, Alresford, Hampshire. 8 m E of Winchester on A31. SU 5833 (OS 185). Mid-Georgian country house, built c. 1750 by Admiral Lord Rodney (d. 1792). The grounds offer a variety of catering arrangements, from picnicking and picking fruit to receptions. Open May to Sept W-Su pm; other times by appt. ⊖ 🅿 ♠ 🍴 🏕

Avington Park, Winchester, Hampshire (tel Itchen Abbas [0962 78] 202). 2½ m NE of Winchester on A33, turn E onto B3047. SU 5332 (OS 185). Charming brick-built country house mainly erected in the late 17th century in the style of Sir Christopher Wren, and enhanced by the Duke of Chandos a century later. The rooms range in style from the early 18th century to the large mid-Victorian conservatory. Avington Church, in Georgian style, stands in the grounds. Open May to Sept S, Su and Bank Hol 1430-1730; open for parties by appt at other times. ⊖ 🅿 WC 🅰 (ground floor only) ♠ 🚻 (by appt) D (on lead, grounds only) ✹ ✗ (compulsory)

Basing House, Old Basing, Hampshire (tel Basingstoke [0256] 467294). 3 m E of Basingstoke, off A30. SU 6652 (OS 185). Ruined Tudor palace built by Henry VIII on the site of four castles. It was an important stronghold for the Royalists during the Civil War, holding out in a two-year siege until 1645. There is a 16th-century gate-house and tithe barn. Open June to Aug daily pm; Apr, May and Sept S, Su and Bank Hols pm. 🚻 🍴

Bishop's Waltham Palace, Bishop's Waltham, Hampshire (tel Bishop's Waltham [048 93] 2460). 6 m E of Eastleigh on A333. SU 5417 (OS 185). Fortified palace dating from the reign of King Stephen, and built by the Bishop of Winchester. William of Wykeham died here in 1404. It has a great tower, gatehouse and Romanesque chapel. Open daily am and pm; Su pm only. ⊖ 🅿 WC 🚻 D (on lead, grounds only) ♠ ✹ EH

Mottisfont Abbey, Mottisfont, Hampshire (tel Lockerley [0794] 40757). 4½ m NW of Romsey off A3057. SU 3226 (OS 185). Founded as an Augustinian priory in the 12th century, it was converted into a private house by Lord Sandys in the 1540s and altered again in the 18th century when the south front was added. There is a room with Rex Whistler *trompe l'oeil* murals, a walled garden with a collection of old roses, and a large and beautiful park. The stable block contains a local history museum. House open Apr to end Sept W and S pm; grounds open T-S pm. 🅿 WC ☐ (by appt) ♿ ♣ ◆ NT

Titchfield Abbey, Titchfield, Hampshire (tel Fareham [0329] 43016). 3 m W of Fareham on A27, turn N. SU 5306 (OS 196). 13th-century abbey converted into a mansion in the 1540s after the Dissolution of the Monasteries by Thomas Wriothesley, the future Earl of Southampton. Much of the Tudor house was destroyed in the 1780s, but the gatehouse remains and the plan of the abbey can be traced in the ruins. Open daily am and pm (Su pm only). ⊖ 🅿 WC ♿ ☐ D (on lead, grounds only) ♣ ◆ ⚘ EH

Isle of Wight

Appuldurcombe House, Wroxall, Isle of Wight (tel Ventnor [0983] 852484). 3 m N of Ventnor on B3327. SZ 5479 (OS 196). Ruins of an 18th-century mansion built for the Worsley family from 1701 in Classical style. The house was irrevocably damaged in 1943 and only the shell of the house remains (though the stonework is in good condition), as does the park, which was the work of Capability Brown. Open daily am and pm (Su pm only). ⊖ 🅿 WC ♿ (uphill from car park) ☐ D (on lead, grounds only) ♣ ◆ ⚘ EH

Arreton Manor, Arreton, Isle of Wight (tel Newport [0983] 528134). 2 m SE of Newport on A3056. SZ 5386 (OS 196). 17th-century manor house with contemporary furniture. It contains a collection of lace-making, and a museum of child-hood. In the grounds (and connected to the house by a secret passageway) is the Pomeroy museum dolls house, and a collection of wireless receivers

dating back to the early 20th century. Open Easter to end Oct, M-F am and pm, Su pm only. ⊖ 🅿 WC ☐ D ♣ ♥ ☶ ◆ ⚘

Haseley Manor, Hazeley Combe, Arreton, Isle of Wight (tel Isle of Wight [0983] 865420). 3 m SE of Newport on A3056. SZ 5485 (OS 196). 14th-century house modified in the late 19th century. The dining room has a mid-Tudor fireplace. There is a small museum of rural life, pottery demonstrations, and children's play area. Open daily am and pm (closed 25 Dec).⊖ 🅿 WC ♿ ☐ D ♣ ♥ ☶ ◆ ⚘ 🕯 🎎

Morton Manor, Brading, Isle of Wight (tel Sandown [0983] 406168). 1½ m N of Sandown off 3055. SZ 6085 (OS 196). Restoration house, built in 1680, and with period furniture. The gardens are terraced and landscaped attractively. Open 1st Su Apr to end Oct daily (exc S) am and pm. ⊖ 🅿 WC ♿ (limited access) ☐ (by appt) D ♣ ♥ ◆ 🕯 🎎

Norris Castle, East Cowes, Isle of Wight (tel Cowes [0983] 293434). ½ m E of East Cowes off A3021, on coast. SZ 5196 (OS 196). Fantasy castle built in 1799 by James Wyatt, a childhood resort of Queen Victoria (who later had Osborne House built nearby).The central tower houses the ballroom. The castle contains 18th- and 19th-century furniture. Open for parties only, by appt. 🅿 WC ♿ (limited) ♣ ♥ (by appt) ☶ ⚘ 🕯

Nunwell House, Brading, Isle of Wight (tel Isle of Wight [0983] 407240). 3 m S of Ryde on A3055, turn W. SZ 5987 (OS 196). Tudor and Jacobean house, with later additions. It contains period furniture in the Jacobean and Georgian wings, and houses a unique collection of family militaria. Open Whit Su to Sept Su-Th pm. ⊖ 🅿 WC ☐ (by appt) ♣ ♥ ☶ ◆ ⚘ 🕯 ● 🍴

Kent

Chartwell, Westerham, Kent (tel Edenbridge [0732] 866368). 2 m S of Westerham on B2026, turn SE. TQ 4651 (OS 188). 18th-century farm-house modified in the 1920s, the home of Sir Winston Churchill from 1924 until his death in 1965. The interior is partly preserved as it was in his life-time, and also contains a museum of Churchill memorabilia; the grounds are attractive and contain a lake, and Churchill's studio hung with many of his paintings. House and gardens open Apr to Oct S, Su, T-Th (also Bank Hol M) late am and pm; house only open Mar and Nov S, Su, W late am, pm. ⊖ 🅟 WC 🅖 🖥 (by appt) D (gardens only) ♣ 🍴 ◆ ✺ 𝐾 (by appt) ● (not in house) NT

Cobham Hall, Cobham, Kent (tel Shorne [047 482] 3371). 4 m W of Rochester, off A2. TQ 6868 (OS 178). House built in the late Elizabethan period, with the central portion completed in the 1660s. Inigo Jones and James Wyatt were both associated with it. The interior contains Elizabethan fireplaces, and the dramatic 17th-century gilt room. The grounds were laid out by Humphry Repton, and the house is now a girls' public school. Open Easter and in school summer holidays, certain days only. 🅟 WC 🖥 (by appt) ♣ 🍴 ⼏ ✺ 𝐾

Eastbridge Hospital, High St, Canterbury, Kent (tel Canterbury [0227] 62395). In city centre. TR 1457 (OS 179). One of the oldest buildings in Canterbury, dating back to the 12th century, and used as an almshouse since the 16th century. Open daily am and pm (exc Christmas and Good Friday). ⊖ 🅟 ★

Finchcocks, Goudhurst, Kent (tel Goudhurst [0580] 211702). 12 m E of Tunbridge Wells on A262, turn S. TQ 7036 (OS 188). Medium-sized brick house built in 1725 for Edward

Bathurst, and containing a collection of historic keyboard musical instruments, including early pianos. Open Easter to Sept Su; also Bank Hols and Aug W-S pm. 🅟 WC 🅖 🖥 (by appt) ♣ 🍴 ⼏ ◆ ✺ 𝐾 ● (no tripods)

Gad's Hill Place, Rochester, Kent. 3 m NW of Rochester, on A226. TQ 7170 (OS 178). Late Georgian house, the home of Charles Dickens from 1858 to 1870, now used as a school. The garden extends through a tunnel under the road. Open by appt only. ⊖ 🅟 ★

Godinton Park, Ashford, Kent (tel Ashford [0233] 20773). 1½ m W of Ashford on A20, turn S. TR 9843 (OS 189). Jacobean house, with a 15th-century great hall and much fine carving and panelling, especially on the main staircase. The rooms contain interesting portraits, china and furniture. There is a formal garden with topiary, laid out in the late 19th century. Open Easter, June to Sept Su and Bank Hol M 1400-1700; other times by appt. ⊖ 🅟 WC 🅖 🖥 (by appt) D (on lead, grounds only) ♣ ● (not in house)

Great Maytham Hall, Rolvenden, Kent (tel Cranbrook [0580] 241346). 4½ m W of Tenterden on A28, turn E. TQ 8430 (OS 188). Georgian-style house designed in 1910 by Sir Edwin Lutyens for the Tennants family. It has now mostly been converted into flats. Open May to Sept W and Th pm. ⊖ 🅟 WC 🖥 (by appt) ♣ ✺ 𝐾 CHA

Owletts, Cobham, Kent. 4 m W of Rochester on A2, turn S onto B2009. TQ 6668 (OS 177). Modest red-brick house built in the 1680s with richly-decorated plasterwork over the stairs. There are small, attractive gardens. Open Apr to Sept W and Th pm. ⊖ 🖥 (by appt) ♣ ● (not in house) NT

Port Lympne, Lympne, Hythe, Kent (tel Hythe [0303] 64646). 3 m W of

Hythe. TR 1034 (OS 179). Early 20th-century house designed by Sir Herbert Baker in Dutch colonial style, and with a number of original features including a tent room by Rex Whistler and a Moroccan patio. There is a wildlife art gallery, gardens, and a zoo park. Open all year daily am and pm (closed 25 Dec). ⊖ 🅿 WC ♿ (limited access) 🚻 ♣ 🍴 ⊓ ◆ ✽ 𝑘 (by appt) 🐕

Quebec House, Westerham, Kent (tel Westerham [0959] 62206). 5 m W of Sevenoaks on A25. TQ 4454 (OS 187). Gabled 16th- and 17th-century red-brick house, the childhood home of General James Wolfe who captured Quebec from the French in 1759. The house and coach house contain an exhibition of pictures and objects related to Wolfe. Open Apr to Oct M-W, F, Su pm; March Su pm only. ⊖ 🚻 (by appt) ♣ ✽ ● (not in house) NT

Quex Park, Birchington, Kent (tel Thanet [0843] 42168). 5 m W of Margate on A28, turn S onto B2048. TR 3168 (OS 179). 19th-century house containing Major Percy Powell-Cotton's ethnographic and natural historical objects from Africa and Asia, collected between 1887 and 1938. These include dioramas, stuffed animals, firearms, *objets d'art*, costumes etc. There are other rooms with 18th-century furniture, and an extensive park. Open Apr to Sept, W, Th and Su pm; also F in Aug and Bank Hols; parties at other times. Museum only open Oct to Mar Su pm. ⊖ 🅿 WC ♿ (limited access) 🚻 (by appt) ♣ 🍴 ◆ ✽ ●

Riverhill House, Sevenoaks, Kent (tel Sevenoaks [0732] 452557). 3 m S of Sevenoaks on A225, turn E. TQ 5452 (OS 188). Small country house, with a notable and varied garden, including terraces, rhododendrons and azaleas. Garden open Apr to Aug Su and M 1200-1800; house (adults only) Bank Hol Su, M 1400-1800. 🅿 WC 🚻 (by appt) ♣ 🍴 ⊓ ◆ ✽ 𝑘 (by appt, for parties) ● (not in house)

Smallhythe Place, Tenterden, Kent (tel Tenterden [05806] 2334). 3 m S of Tenterden on B2082. TQ 8930 (OS 189). Half-timbered 16th-century yeoman's house in a medieval shipping village; it was the home of actress Ellen Terry from 1899 to 1928, and contains an extensive collection of memorabilia and objects associated with the theatre in the early 20th century. Open March S and Su pm; Apr to Oct daily (exc Th and F) pm. ⊖ 🚻 (by appt) ♣ ✽ ● (not in house) NT

Stoneacre, Otham, near Maidstone, Kent (tel Maidstone [0622] 861861). 3 m SE of Maidstone on A20, turn S. TQ 8053 (OS 188). Small half-timbered 15th-century manor house, restored in the 1920s, and with a crown post roof. The small gardens are attractive. Open Apr to Sept W and S pm. 🚻 ♣ ✽ 𝑘 ● (not in house) NT

Temple Manor, Knight Road, Strood, Rochester, Kent (tel Medway [0634] 78743). ½m S of Strood centre. TQ 7368 (OS 178). 13th-century house built for the officers of the Knights Templar, with 17th-century extensions. The house is now empty of furnishings, but has a fine hall and vaulted undercroft. Open Apr to Sept M, W, Th, S am and pm; Tu am only, Su pm only. 🅿 ♿ (grounds only) 🐕 (on lead, grounds only) 🚻 ♣ ✽ EH

Tudor House, King Street, Margate, Kent (tel Thanet [0843] 25511 ext 217). In town centre. TR 3570 (OS 179). Timber-framed house of the early Tudor period, restored in 1951. The plaster ceilings and exposed timberwork of the interior are notable, and the house also contains a local history museum and a collection of sea shells. Open May to Sept M-S am and pm. ⊖ 🅿

Greater London

Apsley House, Hyde Park Corner, London SW1 (tel 01-499 5676). At Hyde Park Corner. TQ 2879 (OS 176). Home of the Duke of Wellington from 1817, and built in 1771-78 by Robert Adam for Lord Bathurst. It was given its present Classical façade in 1828. It was converted into a Wellington Museum in 1947, and the interior has been redecorated in early 19th-century style. It contains fine porcelain and silver, as well as the Duke's collection of paintings, which includes works by Velasquez, Van Dyck and Goya. Open T-Th, S 1000-1750; Su 1430-1750 only. Closed Christmas and New Year. ⊖ WC ☒ (limited, by appt) ☐ ◆ ⚹ ⚔ (by appt, phone 01-589 6371: V&A Education Dept)

Banqueting House, Westminster (tel 01-930 4179). On Whitehall. TQ 3080 (OS 176). One of the first Classical-style houses in London, built by Inigo Jones in 1619-22 as part of the vast Whitehall Palace, which was otherwise never completed. It was the centre of court life in the 17th century. The interior is decorated by Rubens, and the house was converted into a royal chapel by Sir Christopher Wren in the early 18th century, and restored in the 1970s. Open daily (exc M) am and pm (Su pm only); may close at short notice. ⊖ ☐ ★ ◆ ⚹ EH

Carlyle's House, 24 Cheyne Row, Chelsea, London SW3 (tel 01-352 7087). N of Chelsea Embankment. TQ 2777 (OS 176). Early 18th-century town house, the home of philosopher Thomas Carlyle from 1834 to his death in 1881. The house remains as it was at that time and contains many books and relics of Carlyle. There is a small garden. Open Apr to Oct W-Su and Bank Hols 1100-1700. ⊖ ☐ (by appt) ◆ ⚹ NT

Dickens House Museum, 48 Doughty St, Bloomsbury, London

WC1 (tel 01-405 2127). TQ 3082 (OS 176). Georgian brick town house, the home of Charles Dickens from 1837 to 1839 where he wrote *Oliver Twist*, *Nicholas Nickleby* and finished *Pickwick Papers*. The house contains many personal and literary relics, and the reception rooms are preserved as they were in his day. Open M-S am and pm (exc Bank Hols). ⊖ ☐ ◆ ◆ ⚹ ⚔ ● (no flash)

Down House, Downe, Greater London (tel Farnborough [0689] 59119). 5½ m S of Bromley, on A233, turn E. TQ 4361 (OS 177). Early 19th-century house, the home of Charles Darwin for 40 years. It contains many objects associated with Darwin's career, and the Darwin memorial gardens. Open March to Jan T-Th, S and Su 1300-1730 (also Bank Hol M). ⊖ (½ mile) ▮ WC ☒ ☐ (by appt) ◆ ◆ ⚹

Eltham Palace, Eltham, London SE9 (tel 01-859 2112). 9 m SE of central London on A20, turn S. TQ 4273 (OS 177). Medieval royal palace extended in the 15th and 16th centuries but little used from the time of Elizabeth I. There is a 15th-century brick great hall with hammerbeam roof built for Edward IV, and other timber-framed buildings. The palace is the headquarters of the Institute of Army Education. The moat is crossed by a medieval bridge. Open Th and Su am and pm. ⊖ ☒ ☐ D (on lead, grounds only) ◆ ★ ⚹ EH

Fenton House, Windmill Hill, Hampstead, London NW3 (tel 01-435 3471). On W side of Hampstead Grove, 4 m N of London centre. TQ 2686 (OS 176). Late 17th-century house set in a walled garden built for a London merchant. It contains a collection of 18th-century porcelain and the Benton Fletcher collection of early keyboard musical instruments, demonstrations of which may be heard. Open Apr to Oct 1100-1700; March S and Su 1400-

1700. ⊖ WC ♿ (by appt) ♣ ⚹ 𝕏
⬤ (no flash) NT

Forty Hall, Forty Hill, Enfield, Greater London (tel 01-363 8196). 11 m N of central London off A10; 1 m N of Enfield. TQ 3398 (OS 176). Mansion built in 1629 for Sir Nicholas Raynton, with 17th- and 18th-century decoration. The house contains an art gallery and furniture museum. It is set in a wooded park. Open daily (exc M) am and pm. ⊖ ♿ WC ♿ ♿ ♣ ♥ ♿ ★ ◆ ⚹ ⚹ 𝕏 (by appt) ⚹

Hogarth's House, Chiswick, London W4 (01-994 6757). 6 m W of central London, off A4. TQ 2177 (OS 176). Brick-built country house residence of the painter William Hogarth from 1749 to 1764, now housing a Hogarth museum including prints of his best-known works. Open daily am (1100) and pm (Su pm only); closed first two weeks in Sept, three weeks over Christmas and New Year. ⊖ ♿ ♿ (limited access) ♿ ♣ ★ ⚹ 𝕏

Dr Johnson's House, 17 Gough Square, London EC4 (tel 01-353 3745). Off Fleet St. TQ 3181 (OS 176). Large late 17th-century house, the home of Dr Johnson from 1748 to 1759. He compiled his *English Dictionary* and other works in the attic. There is a collection of objects, letters, books and paintings connected with Dr Johnson. Open daily am and pm (exc Su, Bank hols, 24 Dec). ⊖ ♿ ◆ ⚹

Keats' House, Well Walk, Hampstead, London NW3 (tel 01-435 2062). S end of Hampstead Heath. TQ 2785 (OS 176). Early 19th-century house, the home of John Keats from 1818 to 1820. It was restored in the 1970s and contains a collection of objects and papers connected with the poet and his fiancée Fanny Brawne. Open exc Christmas, New Year, Easter F-S and May Bank Hol, daily am and pm

(Su and Bank Hols pm only). ⊖ ♿ WC ♿ (by appt) ♣ ★ ◆ ⚹ 𝕏 (by appt)

Linley Sambourne House, 18 Stafford Terrace, London W8 (tel 01-994 1019). N of Kensington High St. TQ 2579 (OS 176). 19th-century house, the home of Linley Sambourne, political cartoonist for *Punch* magazine in the late 19th century. It is decorated in the style of the period and contains paintings and drawings by many artists of that time. Open March to Oct W am and pm, and Su pm only. ⊖ WC ♿ (by appt at other times) ◆ ⚹ 𝕏

Little Holland House, Beeches Avenue, Carshalton, South London. 3 m E of Sutton, on B278. TQ 2763 (OS 176). The home of Arts and Crafts designer Frank Dickinson, built to his own design and featuring his interior decor, painting, handmade furniture and other craft objects expressing his philosophy and theories. Open March to Oct 1st Su in month and Bank Holo Su and M pm. ⊖ ♿ ★

Marlborough House, Pall Mall, London SW1 (tel 01-930 9249). W of Trafalgar Square. TQ 2980 (OS 176). House built for Sarah, Duchess of Marlborough in 1709-11 by Sir Christopher Wren, and decorated with heroic paintings of Marlborough's battles by Laguerre. It was used as a royal residence from 1817 to 1953, and in the 1960s it became a Commonwealth conference and research centre. Open M-F when not in use for conferences, by appointment only. ⊖ ♿ ♿ ⚹ 𝕏 ⬤

Ranger's House, Blackheath, London SE10 (tel 01-853 0035). On S side of Blackheath, 6 m SE of central London on A2, turn N. TQ 3876 (OS 177). Red-brick late 17th-century house, the home in the mid-18th century of the Earl of Chesterfield, who added the bow-windowed gallery. The house is now an art gallery, with a collection of musical instruments. Open daily am and pm. ⊖ ▮ WC ⊠ ⊟ ♣ ★ EH

Sir John Soane Museum, 13 Lincoln's Inn Fields, London WC2 (tel 01-405 2107). ½ m N of Aldwych. TQ 3081 (OS 176). The home of architect Sir John Soane, builder of the Bank of England. He designed it in 1812, and it now houses his extensive and varied collection of antiquities and paintings, as well as his architectural drawings.Open T-S 1000-1700. ⊖ WC ⊠ (limited access) ⊟ (by appt) ★ ◆ ☀ ✗ (by appt) ● (no flash)

Wesley's House, 47 City Road, London EC1 (tel 01-253 2262). 1 m NE of St Paul's. TQ 3282 (OS 176). Late 18th-century town house, the home of John Wesley at the time of his death in 1791. It contains an extensive collection of objects and papers related to him. Next door is John Wesley's Chapel, opened in 1778 and restored in 1978. It houses a library on the history of Methodism. Wesley is buried in the graveyard. Open daily 1000-1600 (exc 25-26th Dec). ⊖ WC ⊠ (limited access) ⊟ (by appt) ♣ ♥ ☀ ✗ (by appt)

Boston Manor, Boston Manor Road, Brentford, Middlesex (tel 01-570 7728 ext 3974). Just N of A4, 2 m E of Hounslow. TQ 1678 (OS 176). Jacobean house dated 1622, with an elaborate plasterwork ceiling, and original fireplace and mantelpieces. The oak staircase is memorable. Open May to Sept S 1400-1630. ⊖ ▮ WC ⊟ (by appt; tel 01-994 1008) ★

Orleans House, Riverside, Twickenham, Middlesex (tel 01-892 0221). 11 m W of central London, ½ m E of Twickenham. TQ 1673 (OS 176). 18th-century house, of which the only original surviving part is the octagon, built by James Gibbs in 1720, with exquisite decorative plasterwork. The house was the residence of Louis Philippe, Duc d'Orleans, during his period of exile, from 1830 to 1848. There is an art gallery in the house, which is set in woods. Open daily (exc M) pm. ⊖ ▮ ⊠ ♣ ★ ◆ ☀

Surrey

Albury Park, Albury, near Guildford, Surrey (tel Shere [048 641] 2694). 6½ m E of Guildford, on A25. TQ 0647 (OS 187). Country mansion, built in the late 17th century, modified by Sir John Soane, and with the façade designed by A.G.W. Pugin in the mid-19th century; every chimney is different. The interior contains a staircase by Soane. Open May to Sept W and Th pm. ⊖ (1½ miles) ▮ WC ⊟ ♣ ✗ NT

Clandon Park, West Clandon, near Guildford, Surrey (tel Guildford [0483] 222482). 3 m NE of Guildford on A247. TQ 0451 (OS 186). Palladian house built in the early 1730s for the Onslow family. It contains one of the finest 18th-century interiors in Britain, recently decorated in original style. There is a Baroque entrance hall, and a collection of fine 18th-century furniture, ceramics and paintings. The Victorian kitchen is on display. There is also a collection of Chinese porcelain birds, and the museum of the Queen's Royal Surrey Regiment. There is a garden containing a Maori house, and a landscaped park (not open to the public). Open Apr to mid Oct daily exc M, F 1400-1800; Bank Hol Su and M 1100-1800. ▮ WC ⊠ ⊟ (by appt) ♣ ♥ ⋒ ◆ ⊛ ✗ (by appt) NT

Claremont House, Esher, Surrey (tel Esher 67841). 5 m SW of Kingston-upon-Thames on A307, turn W. TQ 1363 (OS 176). Palladian-style house built in 1760s by Henry Holland, Sir John Soane and Capability Brown for Robert Clive of India. The house is now used as a school. It is set in the earliest landscaped park in Britain. This was originally laid out by Vanbrugh and Bridgeman, then modified by William Kent before being reworked by Brown. There is a Gothick belvedere tower built by Vanbrugh in 1717. House open Feb to Nov, first S and Su in the month pm; gardens daily am and pm (exc Christmas). ⊖ (1½ miles) ▣ WC ⊟ ♣ ▬ ⋒ ⋇ 𝑲 NT (grounds only)

Guildford House, 155 High St, Guildford, Surrey (tel Guildford [0483] 505050 ext 3531). In town centre. SU 9950 (OS 186). Timber-framed Restoration period house with fine carved staircase, ironwork and plasterwork. The house now contains a series of temporary exhibitions. Open M-S am and pm (closed if preparing exhibitions). ⊖ ▣ (limited) ⊟ ★ ◆ ⋇ 𝑲 (by appt)

Hatchlands, East Clandon, Surrey (tel Guildford [0483] 222787). 5 m NE of Guildford, on A246. TQ 0652 (OS 187). Mid 18th-century brick house built for Admiral Boscawen. The interiors are an early example of the work of Robert Adam, with later modifications. There is an attractive garden. Open Apr to mid-Oct W, Th and Su pm; also Bank Hol M pm. P ⓦⓒ ▣ ⊟ (W and Th) ♣ ▬ NT

East Sussex

Anne of Cleves' House, Southover High St, Lewes, East Sussex (tel Lewes [0273] 474610). On SE side of Lewes, on A275. TQ 4109 (OS 198). A late medieval house with a varied façade of 1530. The house was part of the divorce settlement of Ann of Cleves from Henry VIII. It now houses a museum of local history, dealing particularly with the ironmaking industry of the Sussex Weald, and there are also collections of children's toys and Victoriana. Open mid Feb to mid Nov M-S am and pm; Su pm. WC ⊟ ♣ D ⋒ ◆ ⋇

Bateman's, Burwash, East Sussex (tel Burwash [0435] 882302). 8 m NE of Heathfield, off A265. TQ 6723 (OS 199). Gabled 17th-century house built for a local ironmaster, home of Rudyard Kipling from 1902 to 1936. The ground floor has 17th-century furnishings, and upstairs is Kipling's study preserved as he left it. A watermill adapted by him to supply electricity is nearby. Kipling's Rolls Royce is in the garage. Open Apr to end Oct daily exc. Th, F 1100-1800 (open Good Fri). ▣ WC ⊟ (by appt) ♣ ▬ ◆ ⋇ ● (not in house) NT

Beeches Farm, near Uckfield, East Sussex (tel Uckfield [0825] 2391). 10 m E of Haywards Heath, off A272. TQ 4520 (OS 198). A 16th-century timber-framed and tile-hung farm-house; the gardens and views are notable. Garden open daily am and pm, house open by appointment only. ⊖ ▣ ⊟ (by appt) ♣ ▬ (by appt) D (on lead) ◆ 𝑲 ●

Brickwall House, Northiam, Rye, East Sussex (tel Northiam [07974] 2494). 7 m NW of Rye off A268. TQ 8323 (OS 199). Gabled Jacobean house, home of the Frewen family since 1666. The drawing room is decorated in original style, and the garden is in the process of restoration in the 18th-century style. The house is now used as a school. Open end Apr to late May, and June to end Sept, W and S 1400-1600. ⊖ ▣ ⊟ (by appt) ♣ ● (by arrangement)

Clergy House, Alfriston, East Sussex (tel Alfriston [0323] 870001). 5 m N of Eastbourne on A22, turn W onto A27 for 4 m then turn S. TQ 5103 (OS 199). Timber-framed and thatched cottage of about 1350. It has a great hall, solar and store rooms. This was the first building to be acquired by the National Trust, in 1896; the house contains an exhibition of life in Chaucer's England. Open Apr to Oct daily am and pm. ⊖ ▣ ⊟ ♣ ◆ ⋇ ● (not in house) NT

Glynde Place, near Lewes, East Sussex (tel Glynde [079 159] 248). 6 m SE of Lewes on A27, turn N. TQ 4509 (OS 198). Large 16th-century house, with bow windows and gables, and built arund a courtyard. The interior was modified in the mid-18th century, and there are many paintings, including portraits by Lely, Kneller and Zoffany of members of the Brand family. Open June to Sept W and Th pm; also Easter Su and M and Bank Hols. ⊖ ▣ WC ⊟ (by appt) ♣ ▰ ⁂ ✗ (by appt) ●

Great Dixter, Northiam, East Sussex (tel Northiam [07974] 3160). 8 m W of Rye, off A28. TQ 8125 (OS 199). 15th-century half-timbered hall manor house, modified by Sir Edwin Lutyens in 1910, attaching another hall to the main body of the house. The great hall has an unusual roof, combining hammer-beams with crown post construction. The gardens were laid out by Lutyens, with a wide range of plants and flowers. Open Apr to mid-Oct daily (exc M) pm; gardens also open 1100 on Spring Bank Hol S, Su and M, and Aug Su. ⊖ ▣ WC ⊟ (by appt) ♣ ▨ ✗ ● (not in house)

Haremere Hall, Etchingham, East Sussex (tel Etchingham [058 081] 245). 8 m N of Battle, off A265. TQ 7226 (OS 199). Early 17th-century manor house, with panelled great hall and period furniture. There is a collection of oriental *objets d'art*. The grounds include terraced gardens, and a shire horse farm, with daily demonstrations. Grounds open Easter to end Sept daily am and pm; house open Bank Hol weekends pm, parties at other times by appt. ⊖ ▣ WC ⊟ (by appt) ♣ ▰ ⊟ ◆ ✗ ● (not in house) ⊀ ⋀

Lamb House, West St, Rye, East Sussex. In town centre. TQ 9220 (OS 189). Early 18th-century red-brick house built for James Lamb, and the home of writer Henry James from 1898 to 1916. The study where many of his greatest novels were written was destroyed during the Second World War, but the house remains furnished in the style of his day. The gardens are attractive. Open Apr to Oct W, S 1400-1800. ⊖ ⊟ (by appt) ♣ ⁂ ● (not in house) NT

Monk's House, Rodmell, East Sussex. 4 m SE of Lewes, TQ 4105 (OS 198). Small 17th-century village house, the home of Virginia and Leonard Woolf from 1919 until his death in 1969. Open Apr to Oct W and S pm. ⊖ ♣ ⊟ (by appt) ⁂ ● (not in house) NT

Preston Manor, Preston Park, Brighton, East Sussex. 2 m N of town centre on A23. TQ 3006 (OS 198). House built in the 1730s in the Classical style, and extended in the early 20th century. The house contains the Macquoid collection of 16th- and 17th-century furniture. Open W-S am and pm, Su pm only (closed Christmas and Good Fri). ⊖ ▣ WC ⊟ (by appt) ♣

West Sussex

Danny, Hurstpierpoint, West Sussex (tel Hurstpierpoint [0273] 833000). 6 m N of Brighton on A23, turn E onto B2116 for 1 m then S. TQ 2814 (OS 198). Elizabethan E-shaped house, with an early 18th-century façade. Open May to Sept W and Th 1400-1700. ▣ ⊟ (by appt) ♣ ⁂ ✗

Legh Manor, Ansty, West Sussex (tel Haywards Heath [0444] 413428). 4 m W of Haywards Heath, off A272. TQ 2822 (OS 198). Architecturally interesting Elizabethan manor house, with a garden laid out in around 1900 by Gertrude Jekyll. Grounds and Tudor part of house only open, Apr to Oct second and third W and second S in month 1430-1730. ▣ (limited) ▨ (grounds only) ⊟ (by appt) ♣ ⁂

Newtimber Place, Newtimber, West Sussex (tel Hurstpierpoint [0273] 833104). 11 m N of Brighton on A23, turn W. TQ 2613 (OS 198). 17th-century house with a moat and Etruscan-style wall-paintings. Open May to Aug Th pm. ▣ 日 (by appt) ♣ 🗡 ●

Sackville College, High St, East Grinstead, West Sussex (tel East Grinstead [0342] 21639). In town centre. TQ 3938 (OS 187). Quad-rangular Jacobean almshouses, founded in 1609 by Robert Sackville, 2nd Earl of Dorset. The common rooms and hall have their original oak furniture. Open May to end Sept daily 1400-1700. ⊖ ▣ 🖾 日 (by appt) ◆ ⚘ 🗡 (by appt for parties)

Standen, East Grinstead, West Sussex (tel East Grinstead [0342] 23029). 1½ m S of East Grinstead off B2110. TQ 3935 (OS 187). Large house built in the 1890s by Philip Webb, adding to an existing farm-house. The interior retains its original furnishings, including fabrics and wallpaper by William Morris, and other products of the Arts and Crafts movement. The gardens, also laid out by Webb, have fine views across the Medway valley. Open Apr to end Oct W-Su 1400-1800. ▣ WC 日 (by appt) ♣ 戸 ◆ NT

The West Country

HEREFORD AND WORCESTER

POWYS

GWENT

MID GLAMORGAN

SOUTH GLAMORGAN

Cardiff

GLOUCESTERSHIRE

Gloucester

OXFORD-SHIRE

Swindon

Lydiard Park

Horton Court

Dyrham Park

Blaise Castle House

Bristol

Clevedon Court

Red Lodge

The Georgian House

1 Royal Crescent, Bath

Claverton Manor

Bath

AVON

Westwood Manor

King John's Hunting Lodge

Sheldon Manor

1 2

Great Chalfield Manor

Avebury Manor

Bowood House

Littlecote House

WILTSHIRE

Chalcot House

Oakhill Manor

Longleat House

nbe Sydenham Hall

Coleridge Cottage

Barford Park

Gaulden Manor

SOMERSET

Taunton

Poundisford Park

Knightshayes Court

Midelney Manor

Hatch Court

Bishop's Palace, Wells

Lytes Cary

3

5

4

7 6

Stourhead

Sandford Orcas Manor House

Compton House

Yeovil

Purse Caundle Manor

Philipps House

Pythouse

Wardour Castle

Wilton House

Mompesson House

Salisbury

Newhouse

Chettle House

DORSET

HAMPSHIRE

Killerton House

Fursdon House

Cadhay

Exeter

Forde Abbey

Parnham House

Milton Abbey

Kingston Lacy House

Deans Court

A la Ronde

gbrooke

radley Manor

Torre Abbey

Torbay

kham ouse

Wolfeton House

Hardy's Cottage

Athelhampton Hall

Cloud's Hill

Smedmore House

Bournemouth

1 Corsham Court
2 Lacock Abbey
3 Priest's House
4 Tintinhull House
5 Barrington Court
6 Brympton d'Evercy
7 Montacute House

Dyrham Park

near Chippenham, Avon

This mansion house, grouped together with its parish church and stables in a narrow valley, is rather a curiosity; built in two separate stages, it is really two houses rather clumsily joined together. It was built between 1691 and 1702 for William Blathwayt, an ambitious civil servant with a sincere feeling for architecture. He engaged a little-known French architect, Hauduroy, who added a long two-storey front to the old Tudor house. This, the existing west front, joined to the church, has a definite French look, with its projecting wings breaking the monotony of the fifteen bays. By 1694 the work was complete, and Hauduroy disappeared into the obscurity whence he came. Blathwayt became Secretary of State under William III, and the prestigious William Talman was engaged to improve the house further. Talman added what is really a second house, and the great east front, Italianate in style and decorated with carved ornament, was built from 1700 to 1704. The two houses, back-to-back with each other, are joined only by the large hall and dining room. Talman is also credited with a design for the stables, but this rather crude piece of architecture was actually the work of his carpenter, Edward Wilcox. The interior of the house has not altered much since Blathwayt's time, and most of the furniture dates from that period. There is some fine rich decoration, notably in the balcony room, which also contains some of Blathwayt's Delftware vases and has lovely brass door fittings.

☎ Abson (027 582) 2501

6 m N of Bath on A46

ST 7475 (OS 172)

Open Apr, May and Oct: daily exc Th and F (but inc Good Fri); June to end Sept: daily exc F; 1400–1800; other times by appt

🅿 WC ♿ 🚻 D (on lead) ♠ ♣ 🎋 ⚔ ● NT

Antony House

Torpoint, Cornwall

This elegant rectangular house was built for Sir William Carew between 1711 and 1721, of silver-grey stone with rose-covered red-brick wings. The present building has a simple design, with identical north and south façades, nine windows along the first storey, eight below, and dormer windows set into the roof. The front porch was added in 1839. Surprisingly, the interior is quite different from the light and graceful exterior, the rooms downstairs being quite small and intimate, panelled in Dutch oak and consequently rather dark. They contain some fine furniture made for the house, including a Queen Anne pier glass and a set of tapestry-covered chairs, and an excellent series of family portraits. There are also collections of sporting paintings and porcelain; the tapestry room contains Soho tapestries, and the library displays a copy of Richard Carew's *Survey of Cornwall*, published in 1602. The ground-floor rooms, all of which face north and only one with windows in more than one wall, are arranged round a central hall, from which a fine if rather heavy staircase rises in two broad flights. The first floor is much more 18th century in feeling, with charming bedrooms leading off a tall-arched central corridor. The beautiful gardens, which run down to the River Lynher, were partly laid out by Humphry Repton, and many of the trees he planted have grown to an enormous size. The stone carvings and the Burmese temple bell were brought back by Sir Reginald Pole-Carew from military campaigns in India and Burma.

☎ Plymouth (0752) 812191
1 m NW of Torpoint on A374 turn NW at Maryfield.
5 m W of Plymouth via Torpoint car ferry
SX 4156 (OS 201)

Open 1st Apr to 31st Oct T-Th and Bank Hol M
1400-1800

♿ (limited) P ♿ (limited access) 🚻 (by appt) ♣ 🍴 ♦ ✶ 🐾 ● NT

Cotehele House

St Dominick, near Saltash, Cornwall

A romantic granite-built manor house with an authentic medieval atmosphere, very little altered since it was built by Sir Richard Edgcumbe from 1480. The tower was added in 1627. But if the house has not changed much, the setting undoubtedly has. Sir Richard, who was involved in the revolt against Richard III, was a hard-headed soldier; he would neither have recognised nor comprehended the soft lush lawns and vegetation which now surround his house. Woods would never have been allowed so near a house, and gardens were at that time strictly utilitarian. But Cotehele still has tremendous atmosphere, and gives the visitor a sense of stepping back into the past which few houses can achieve. The Edgcumbes more or less left the house after 1700, though they did not neglect it, and tourists, among them George III, had begun to visit it by the end of the 18th century – 'old-fashioned houses' were becoming popular at the time. The great hall is decorated with armour, antlers and oak furniture, while the small, dark, ground-floor rooms are hung with rich tapestries. There are four-poster beds in the bedrooms, and the ground-floor chapel contains an early clock of Sir Richard Edgcumbe's day. However, it is likely that not all the furniture and fittings are actually relics of the old house – some may have come from elsewhere and from antique dealers during the 19th century, when 'Elizabethan' furniture was often skilfully made up from bits and pieces. The estate includes a watermill and a quay on the River Tamar.

☎ Liskeard (0579) 50434

6 m SW of Tavistock on A390 turn S at Drakewalls to Albaston and Newton

SX 4268 (OS 201)

Open Apr to end Oct daily 1100-1800; house only closed F; gardens only open Nov to Mar

♿ 🅿 WC ⓧ (limited access) 🚻 (by appt, not Bank Hols) 🌳 🍽 (limited opening) 🎏 ◆ 🐕 ✕ NT

Pencarrow

Washaway, Bodmin, Cornwall

This well-proportioned stuccoed Palladian house has a particularly exciting approach, through beautiful woodlands and into a drive with massed hydrangeas, rhododendrons and azaleas which suddenly reveals the house. In the 1760s Sir John Molesworth, 4th Baronet, employed an architect from York, Robert Allanson, to rebuild the family house. Allanson added the south and east fronts in the Palladian style, but traces of the original building can still be seen on the west. The first room to be visited is the music-room; this owes much to Sir William Molesworth, the 8th Baronet, who in 1844 engaged the architect George Wightwick to carry out alterations. He extended the room by adding the alcove, and carried out the maple-graining on the walls. He also used old panelling from another house, both here and in the entrance hall, which he turned into a library. Most of the rooms, however, are mid-Georgian and very charming, with good chimneypieces and fine furniture. The house is well known for its large collection of 18th-century paintings including portraits by Reynolds, a landscape by Richard Wilson and a conversation-piece by Arthur Devis with St Michael's Mount in the background. The nursery, upstairs, contains a bedspread showing the flora and fauna of Cornwall, embroidered by the Cornwall Federation of Women's Institutes, as well as several antique dolls dressed in hand-made baby clothes. In the ante-room a lovely 19th-century linen wallpaper patterned with birds and butterflies was found beneath layers of later paper.

☎ St Mabyn (020884) 369

5 m NW of Bodmin on A389 turn NE to Croanford

SX 0471 (OS 200)

Open Easter to end Sept M, T, W, Th, Su 1330-1700; June to mid Sept and Bank Hol M 1100-1700

♿ (limited) 🅿 WC 🅰 🚻 D ♠ ♣ 🎋 ◆ 🎗
🌿 ⚔ (by appt for parties) 🐕 🦮

A La Ronde

Summer Lane, Exmouth, Devon

An unusual sixteen-sided house, built by two maiden ladies in the last years of the 18th century, and one of the lesser-known curiosities among England's houses. The ladies, Jane Parminter and her orphaned cousin Mary, had spent ten years travelling on the Continent, and they designed the house, supposedly based on the church of San Vitale in Ravenna, on their return in 1794. Later they also built a chapel and almshouses nearby. The house, although it appears small, is actually quite spacious: there are twenty rooms within its solid stone walls, those on the ground floor radiating from a central hall 54 foot high. Between the rooms are closets with lozenge windows, and there are many ingenious space-saving devices such as folding seats and sliding doors. The decorations were all done by the ladies themselves, and provide impressive examples of the handicraft skills that were becoming popular at the time. In the hall, rather than San Vitale's mosaics, the whole upper part is encrusted with shellwork and featherwork pictures of birds. There is more featherwork in the drawing room, while elsewhere there are pictures made of sand and seaweed and some of cut paper. All this would probably have been ridiculed earlier this century, but with today's revival of these 18th- and 19th-century crafts the house has a particular fascination for the modern visitor. The original furniture and pictures are still in the house, which is now owned by a descendant.

☎ Exmouth (0395) 265514

In the N outskirts of Exmouth off B3180

SY 0083 (OS 192)

Open Good Fri to end Oct daily; groups by appt throughout year

⊖ P WC ⓦ 🗓 ♣ 🍴 ⛱ ◆ ♒ ⚲ ●

Fursdon House

near Thorverton, Devon

Fursdon is situated at Cadbury amid unspoiled Devon countryside, in the triangle of land formed by Crediton, Exeter and Tiverton. It is a fine example of a family house that has never been a showpiece but has always functioned as the centre of a busy agricultural estate. Walter de Fursdon lived at Cadbury from about 1259; the property has passed down in the male line ever since, and is still owned by the Fursdons. The present Fursdon House probably originated at about the beginning of the 17th century, but most of the structure now visible was the work of Richard Strong, a Minehead builder employed by George Fursdon in 1732. Strong created a long Georgian building with two wings; the library was added to the west wing in 1815, and three years later the façade was rather clumsily ornamented with an Ionic colonnade. There have been numerous minor and major alterations over the centuries, but exploitation of Fursdon's non-agricultural appeal (for example, by the creation of holiday flats) is quite recent; and the house was only opened to the public in August 1982. The interior contains family portraits, antique furniture and a collection of interesting domestic items in the so-called billiard room (actually a kitchen). There is also a costume collection, including a beautifully preserved mid 18th-century mantua, or court dress, worn by Elizabeth Fursdon. The gardens, parkland and woods are one of the pleasant features of Fursdon, and the walks laid out on the estate include one to the Iron Age hill fort of Cadbury.

☎ Exeter (0392) 860860

6 m SW of Tiverton on A3072, turn S to Fursdon

SS 9204 (OS 192)

Open Easter Su and M; May to end Sept Th, Su and Bank Hol M also W in July and Aug

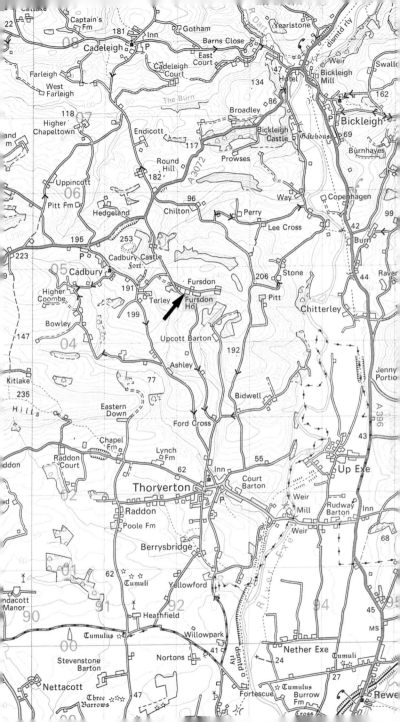

Saltram House

Plympton, Plymouth, Devon

Saltram, a large, square Georgian house, its brickwork rendered and whitewashed, can only be described as plain, but inside these very ordinary walls lie the most out-of-the-ordinary rooms imaginable. The original house was Tudor, and in the 1740s John Parker and his wife Lady Catherine set about enlarging it. The main south front was built at this time; the porch was a Regency addition. This part of the building, and the rooms within it, is Rococo in style, and the charming interiors contain some lovely plasterwork. In 1768 John Parker died, and his son of the same name called in Robert Adam to fit up the east range, which had not been finished. The dining room and saloon were completely remodelled by Adam, down to the last detail, and represent some of his finest work. The saloon, or ballroom, is on the grand scale that was fashionable at the time, and occupies the whole central part of the wing. The dining room, its long, low proportions the result of an earlier phase of building, is pale green and white, with a splendid ceiling the design of which is reflected in the specially woven carpet. Adam's work did not extend to the first-floor rooms, but several of these are quite delightful, and include a Chinese Chippendale room and another which is hung with Chinese paintings on glass. There are also some beautiful and unusual oriental wallpapers in several rooms. All the furniture and paintings, including fourteen by Reynolds, belong in the house, and the Adam rooms are carefully preserved from the harmful effects of light.

☎ Plymouth (0752) 336546

3 m E of Plymouth on A379 turn NE for 1½m at Billacombe and NW to Saltram House

SX 5255 (OS 201)

Open Apr to end Oct Su-Th, Bank Hol F, S 1230-1800; garden only daily throughout year 1100-1800

♿ (¾ m) 🅿 WC ♿ (by appt) 🚻 D (grounds only)

♣ 🍴 ◆ ✾ ● (not in house) ⚘ NT

Athelhampton Hall

Athelhampton, Dorset

In spite of partial destruction and much alteration and restoration, Athelhampton is an important example of Tudor domestic architecture on a comparatively modest scale. The name indicates Saxon origins. The house itself dates from the last years of the 15th century, when Sir William Martyn built a typical early Tudor country house, with towers and ornamental battlements, round a small court. The two-storey wing on the west side was added by his son, and his grandson built an outer courtyard to the south-west, which was later demolished. After the death of the last male Martyn in 1596 the estate was divided, and passed through various hands until it was acquired in 1848 by George Wood. By this time only the foundations remained of the north and east wings round the small court, and the rest of the house was becoming derelict. Wood pulled down the decaying gatehouse and walls of the outer courtyard, leaving the open front as it exists today. He began the restoration of the house, most importantly saving the superb, simple roof of the Great Hall, which also contains fine linenfold panelling. Restoration continued under Alfred Cart de Lafontaine later in the 19th century. He also began to create the gardens, which have recently been extended and improved, on a formal Italian pattern. There are still formal gardens, divided by walls of Ham stone into four separate plots, as well as woodland and riverside walks and an early 16th-century dovecot.

☎ Puddletown (030 584) 363

5 m NE of Dorchester on A35, 1 m E of Puddletown

SY 7794 (OS 194)

Open W before Easter to 2nd Su Oct W, Th, Su, Good Fri, Bank Hol M, M and T in Aug 1400-1800

⊖ Ⓟ WC ⓚ (limited access) 🚻 (by appt) D (car park only) ♣ ♣ ♦ ⚶

Forde Abbey

near Chard, Dorset

Originally a Cistercian monastery until it was closed by Henry VIII, the house, with its long, low front in the lovely Ham Hill stone, is a fascinating mixture. Unusually, at Forde a great deal of the original monastic building survives, forming the base for the major rebuilding in the 17th century. Thomas Chard, the last abbot, had spent large sums of money on new building just before the Abbey was closed: the great hall and the magnificent gatehouse, both resplendent with Renaissance carving, were built by him. In 1539 he and his twelve monks surrendered the Abbey, and 110 years later it was bought by Edmund Prideaux, Attorney-General to Oliver Cromwell, who set about transforming it into an Italian-style palace. Its monastic layout was well suited to this treatment, and there was little structural alteration, but the interiors were quite transformed, with lavish panelling and plaster-work. The outside was 'modernised' with large mullioned windows (later 'Gothicked'), a saloon was built next to the entrance tower, as well as some small rooms over Abbot Chard's cloister. The alterations were more or less complete by 1660, and little has since been changed. The great hall, which was shortened by Prideaux, has a late Gothic wooden roof and 17th-century panelling, and all the rooms have been furnished in a style suited to the building. The saloon contains five Mortlake tapestries woven from Raphael cartoons, a later present from Queen Anne. A conservatory has been made from the surviving cloister range.

☎ South Chard (0460) 20231

4 m SE of Chard on B3162 turn S at Whatley to bridge

ST 3505 (OS 193)

Open May to Sept Su, W, Bank Hol M 1400-1800

🅿 WC 🚻 (limited access) 🚽 D 🍴 🍷
🪑 ◆ 🎱 ✳ 🏹 (by appt)

Kingston Lacy House

near Wimborne, Dorset

This is one of the earliest surviving houses built in a Classical style after the Restoration of 1660, with such features as pediments, hipped roofs, dormer windows and cornices. The manor of Kingston Lacy took its name from its medieval owners, the de Lacys, who were Earls of Lincoln. It belonged to members of the royal family for long periods until 1637, when Sir John Bankes bought it from the Earl of Newport. Until 1643 the main Bankes family residence was Corfe Castle, which Lady Mary Bankes defended vigorously for six weeks against Parliament during the Civil War. As a result the castle was thoroughly slighted, and after the Restoration Lady Mary's son, Sir Ralph Bankes, had a new family seat built at Kingston Lacy in the mainly urban style that was becoming fashionable. The architect was Inigo Jones's disciple, Sir Roger Pratt, under whom the work was executed between 1663 and 1665. The original house consisted of two storeys and a semi-basement, built in red brick with stone dressings. Kingston Lacy was much altered in the 18th century, but in 1835-39 Sir Charles Barry restored the cupola and other features; however, he also 'improved' it in various ways, for example adding the present chimneys and lowering the ground level to convert the basement into a full floor on the main front. The interior was lavishly remodelled in Barry's favoured Renaissance style, the *pièce de resistance* being the splendid marble staircase. The formal garden is also Barry's creation. The collection of paintings at Kingston Lacy is famous.

☎ Wimborne (0202) 883402

2 m NW of Wimborne on B3082

ST 9701 (OS 195)

Open Apr to end Oct daily exc Th and F; park 1200-1800, house 1300-1700

⊖ Ⓟ WC ⊟ (by appt only) ♣ ♠ ● ⊓ ◆ ☀ NT

Parnham House

Beaminster, Dorset

For 300 years the family seat of the Strodes, Parnham was largely built about 1550, and it retains a charmingly Tudor appearance, with gables and pinnacles and mellow golden stone. But in fact there were three main periods of building, as the whole right half of the house was built about 1600, while in 1810 John Nash was commissioned to enlarge the house. The castellations, buttresses and pinnacles, together with the other two sides of the house, were added at this time. Extensive renovations were carried out in 1910 by Hans Sauer, who reinstated the Tudor interiors altered by Nash and landscaped the grounds, building terraces, water channels, balustrades, rotundas and the front court. Parnham has had a chequered history and, like other historic houses, has several times been under threat. In 1896 it was owned by Vincent Robinson, who lived surrounded by pets and Renaissance furniture; in 1912 it was bought by the Rhodes-Moorhouse family, whose son William was the first pilot to be awarded the Victoria Cross; in the 1920s it became a country club; in the Second World War it was an army hospital and afterwards it became a private nursing home. In 1976 it was bought by John Makepeace, and it is now the home of the John Makepeace Furniture Workshop and the School for Craftsmen in Wood. The main rooms now display, not only their original carving and panelling, but also fine examples of modern furniture, and visitors can see young craftsmen at work in the workshop.

☎ Beaminster (0308) 862204

7 m N of Bridport on A3066 just before Beaminster

ST 4700 (OS 193)

Open Apr to Oct Su, W, Bank Hols; also parties T, Th 1000-1700

⊖ 🅿 WC ♿ 🚻 D (on lead, grounds only) ♠

🍽 ⛲ ◆ ⚒ 🍴 (for parties on T, Th)

Wolfeton House

Dorchester, Dorset

A grey, oddly shaped, mainly Elizabethan manor house is the seat of the Trenchards, a Wessex landowning family who made their money in sheep farming. It was built by two members of the family, Sir Thomas and his great-grandson Sir George. The gatehouse, which has an inscription dating it to 1534, was Sir Thomas' work, and once led to a courtyard house also built by Sir Thomas on the site of a yet earlier house. Unfortunately, most of this house was demolished around 1800, and even the hall was removed. The taller range has large mullioned windows reminiscent of Longleat; it was built by Sir George around 1580; Thomas Hardy described the house as 'an ivied manor house . . . more than usually distinguished by the size of its mullioned windows'. The interior shows a mixture of decorations, but there are some splendid fireplaces, and a vaulted corridor, stone staircase and doorway. All are beautifully carved in Renaissance style, possibly the work of one of the Longleat carvers. When the last Trenchard of Wolfeton departed, taking with him the armorial glass and probably other treasures, the house lost much of its status and was sold and partly demolished, although some restoration was attempted in the mid-19th century. Since the Second World War, the house has been let as three separate residences, but the present owner is making many improvements, and restoration work is in hand. The outbuildings contain a cider house, where old presses are still used, and visitors given a glass of home-made cider on leaving.

☎ Dorchester (0305) 63500

2 m N of Dorchester on road to Charminster

SY 6892 (OS 194)

Open May, Jun, Jul, Sept T, F, Su, Bank Hol M; Aug daily exc S 1400-1800; at other times by appt

⊖ (1½ m) 🅿 🅰 (limited access) 🚻 🌳
🐾 (by appt) ⚹ 𝄞 (by appt)

Brympton d'Evercy

Yeovil, Somerset

This house, with all its buildings made of the lovely golden Ham Hill stone, seems at first sight to be uniformly Tudor. But although the west front, which bears the arms of Henry VIII, is a fine example of Tudor architecture, the house went through several different periods of building. The main house, together with the left side of the front, was built by the Sydenhams from about 1520; the central section, church and dower house are older. The Sydenhams lived consistently above their income, and the programme of constant building and renovation they carried out at Brympton, culminating in the splendid south front, was to turn them from wealthy landowners to paupers. The south front is one of the most famous examples of the Classical style of the 1670s, and was the work of Sir John Posthumous Sydenham; but it finally destroyed the family fortunes. Five years after his death in 1692 his son put the house up for sale as 'a Very Large New Built Mansion', and after a short spell of ownership by Thomas Penny, Receiver General of Somerset, it was bought by Francis Fane, whose descendants still live here. The main ground-floor rooms are open to the public, including the hall with its huge staircase, and they contain some good furniture and paintings as well as 18th-century panelling. The layout of the gardens was the work of Lady Georgiana Fane, who never married because she lost her heart to a 'lowly soldier' and her father would not countenance such a match. The 'lowly soldier' in question later became better known as the Duke of Wellington.

☎ West Coker (093 586) 2528

2½ m W of Yeovil on A3088 turn SW to Odcombe
for ½ m then S

ST 5215 (OS 183)

Open Easter weekend then May to end Sept
S-W 1400-1800

⊖ (1½ miles) 🅿 WC 🔄 🚻 ♣ ☕ ⊓ ◆ ☀

Combe Sydenham Hall

Monksilver, Somerset

This pleasant manor house dates back to about 1360, when it was built for the Sydenham family, who occupied it for over 300 years. Standing in a combe, or hollow, surrounded by trees, Combe Sydenham was originally a semi-fortified house, built round a courtyard with towers at each corner; but it was remodelled by George Sydenham in 1850. Like other comfort- and fashion-conscious Elizabethans he reduced the height of the hall and put in an up-to-date staircase; the present house, with its single tower and two wings, is essentially his creation and, appropriately enough, the visitor is greeted by his initials carved on the porch. Notable features of the interior are the Court Room, with its 15th-century tiles, and the Restoration Chamber; in the latter, the windows were designed to show off the great expanses of glass in which the Elizabethans took such pride. One of the curiosities of the house is a meteorite kept in the entrance hall. According to legend, Sir Francis Drake was betrothed to George Sydenham's daughter, and promised to send her a sign while he was away at sea. The voyage was a long one, and Elizabeth Sydenham grew tired of waiting. She was on the point of marrying another man when the meteorite rolled between them, evidently putting an end to the match; at any rate, she did in fact marry Drake in 1585. In front of Combe Sydenham stands a range of buildings including a well-restored medieval gatehouse.

☎ Stogumber (0984) 56284

5 m N of Wiveliscombe on B3188

ST 0736 (OS 181)

Open Easter M to end Oct M-F 1100-1630;
(Apr, May, Oct 1100-1515)

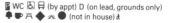 WC 🚻 🚻 (by appt) D (on lead, grounds only)
♣ 🍴 ⛲ ◆ ❀ ● (not in house) ♿

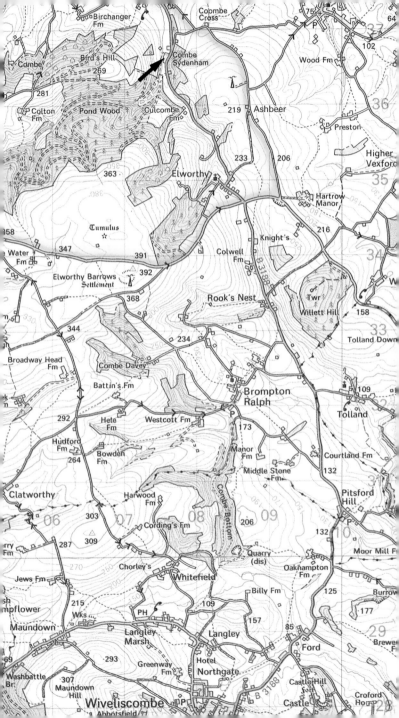

Gaulden Manor

Tollard, Lydeard St Lawrence, Taunton, Somerset

From the outside Gaulden Manor appears to be simply an attractive but unpretentious ancient farmhouse, but its rough stone walls are a front for some of the most extraordinary and sumptuous plasterwork in the country. There is some disagreement about the dates of this work, but it seems that in 1565 the house became the retreat of the ex-Bishop of Exeter, James Turberville, after he had been imprisoned in the Tower of London for refusing to take the Oath of Supremacy to Queen Elizabeth I. Some experts believe that the plasterwork in the hall, by far the most elaborate, was put in by him, while others claim that it is mid-17th century and was put in by his great-nephew John Turberville, who bought the house in 1639. The hall has a deep plaster frieze running right round the walls, with intricately modelled religious scenes (possibly representing the life and trials of the Bishop), and the ceiling has three large plaster roundels, the central one with a pendant. The fireplace is Tudor, and there is a fine mantelpiece bearing the Turberville arms. Opening off the hall, and divided from it by a carved 16th-century oak screen, is a small room known as the chapel, with more plasterwork, and the upstairs bedroom has a fine decorative plaster overmantel which was unfortunately cut in half by a later ceiling. This and the ceiling of the hall were probably done for John Turberville. There is a very attractive garden, which was entirely created by the present owners, who have lived at the Manor since 1966.

☎ Lydeard St Lawrence (09847) 213

8½ m NW of Taunton on A358 turn W to Handy Cross then S for 1 m and N for ½ m

ST 1131 (OS 181)

Open Easter Su and M, 1st Su May to 2nd Su Sept Su and Th; also W July and Aug, and Bank Hol M

🅿 WC ♿ 🚻 (by appt) ♠ 🏪 ♦ 🕸
🌾 𝕏 ● 𝕏 🕯

94

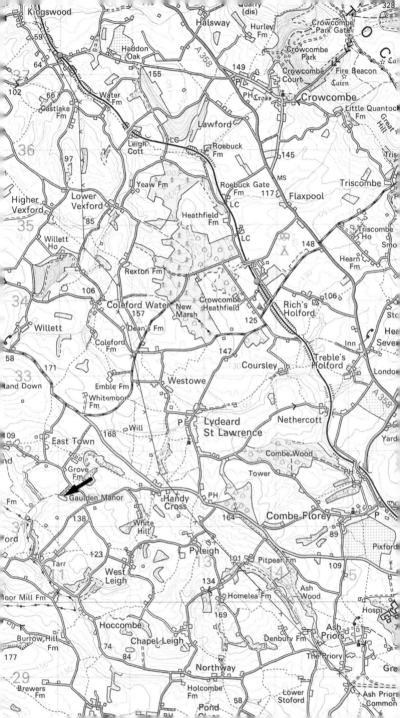

Hatch Court

Hatch Beauchamp, Taunton, Somerset

This elegant 18th-century house was built by John Collins, who had become wealthy in the wool trade and needed a house worthy of his new status. Instead of engaging a professional architect, Collins approached Thomas Prowse, a sophisticated amateur and local gentleman, to provide a design. Prowse was a friend of a more famous amateur architect, Sanderson Miller, who built Hagley Hall near Birmingham, and Hatch Court, though smaller than Hagley, is very similar. The two buildings are exact contemporaries, both being built in 1775. Prowse may have been an amateur, but there is nothing clumsy about Hatch Court, with its four towers, its charming arcaded piazza and its fine, restrained architectural detail beautifully executed in Bath stone. The curving wings, added around 1800, are successful additions, giving the house a satisfying quality. The finest feature of the interior is a marvellous staircase, which fills the entire centre of the house, dividing and doubling back on itself at the half-landing. There is a fine old oak table in the hall, dating from about 1630, and the drawing room has an elaborate Rococo plaster ceiling. The furniture is largely Georgian, and the paintings, as well as family portraits, include several wild Canadian landscapes which belonged to Andrew Gault, who lived here in the 1920s. A dynamic figure, Gault was responsible for raising a Canadian regiment, Princess Caroline's Light Infantry, to fight in the First World War, and a small regimental museum commemorates his achievement.

☎ Hatch Beauchamp (0823) 480208

8 m SE of Taunton on A358 turn NE to Curry Mallet

ST 3021 (OS 193)

Open July to mid Sept Th 1430-1730

⊖ P WC ⊟ (by appt only from May) ♠ ● ⊼ ⋇ 🕇 ● (with permission) Deer park

Montacute House

Montacute, near Yeovil, Somerset

Montacute, built of the honey-coloured Ham Hill stone and glittering with glass, is one of the finest Elizabethan mansions in England, and one of the least changed. It was completed about 1600 for Sir Edward Phelps, a West Country landowner and lawyer, who was Speaker of the House of Commons from 1598, and the architect was probably William Arnold, a brilliant local master-mason. There were some alterations in the late 18th century, when the lovely Renaissance porch now on the west front was brought from Clifton Maybank, and Lord Curzon, who lived here from 1915 to 1925, made some changes to the interior, but the character of the house was unaltered. The interior is also almost pure Elizabethan, and retains much of its original decoration, including fine ornamental plasterwork, carved fireplaces, and a fine carved screen in the great hall. The rooms themselves are large, light and airy, the finest being the library, once the great chamber, where the manorial courts would have been held. There is a splendid chimneypiece of Portland stone, a carved wooden porch, Elizabethan panelling and four great windows. The panelling in the other rooms is 19th century, dating from Lord Curzon's occupancy, and the house has been entirely furnished by the National Trust with the help of a magnificent bequest of furniture and tapestries belonging to Sir Malcolm Stewart. The long gallery shows a large collection of 16th-century portraits from the National Portrait Gallery.

☎ Martock (0935) 823289

4½ m W of Yeovil on A3088

ST 4917 (OS 193)

Open Apr to end Oct daily exc T 1230-1730;
gardens open daily 1230-1730

🅿 WC 🚾 (limited) 🚻 (by appt) D (on lead, grounds only) ♣ 🍴 ⌤ ◆ ✻ ● (no flash) NT

Poundisford Park

Poundisford, Taunton, Somerset

A compact Tudor gabled house built on an H-plan and dating from about 1546, it is unusual for its time in that it stands by itself in the midst of fields instead of being built on a previously fortified or ecclesiastical site. It also still retains its detached kitchen, where spit roasting and brewing would have been done. The house is owned by the Vivian-Neal family, whose home it is, and there are guided tours of the rooms. The entrance door leads into a screens passage and then the great hall, a fine, large room with a 16th-century plaster ceiling. There are several of these in the house, and they are believed to be early examples of the work of a School of West Somerset Plasterers. A spiral stair leads to the porcelain gallery on the upper floor, where there is a ribbed plaster ceiling contemporary with the one in the hall. The glass in the bow window is also 16th century, as are the window fastenings. The gallery contains an English longcase clock in a Chinese-style case, and the cabinets hold china and curios collected over generations. Next comes a bedroom with a fine 19th-century patchwork quilt, and then a room in which 18th- and 19th-century costumes are displayed. A modern staircase leads down to the original pantry, and thence to the pretty 18th-century parlour formed from what was the buttery. The dining room was built on in 1692, but altered in 1737, and is totally 18th century in style. The pleasant, simply planted garden has an unusual 17th-century gazebo and a decorated water-butt bearing the date 1671.

☎ Blagdon Hill (082342) 244

5 m S of Taunton on B3170 turn W

ST 2220 (OS 193)

Open May to mid Sept W, Th and Bank Hol M also F in July and Aug 1100-1700

F WC ♿ (limited access) ⊟ (by appt) ♦ ☙ ♦ ⚹ ⅃
● (not in house)

Corsham Court

Corsham, Wiltshire

Corsham Court, a mixture of Elizabethan and 18th century, is best known for its 18th-century rooms with their fine art collections. The original yellow stone manor was built by Thomas Smythe in 1582, and the south front is unchanged, as are the stables on the west side and the riding school on the east. The house was bought in 1745 by Paul Methuen, and in 1760 Capability Brown, famous as a landscape gardener but less so as an architect, was given the job of enlarging both house and grounds. He converted the entire east wing into a suite of sumptuous state rooms to hold Paul Methuen's collection of works of art. In 1800 John Nash was commissioned to make further enlargements, but his 'Gothick' north front and entrance was replaced in 1845 by Thomas Bellamy, and all that now remains of Nash's work is a tiny dairy in the park and some decoration on the east front. Of the rooms open to visitors three, the entrance hall, music room and dining room, are Bellamy's; the rest are all Brown's creations, the finest of them being the picture gallery. This room is hung with pictures up to the ceiling, and the walls are covered in crimson silk damask echoed in the covers of the chairs, sofas and window seats made by Chippendale. The plaster ceiling is from a rejected design by Brown for the hall at Burton Constable, and the carpet reflects this design. The other state rooms are smaller but no less treasure-filled. The paintings include an *Annunciation* from the studio of Filippo Lippi and works by Elsheimer, Jan Breughel and Teniers, and the furniture is of high quality.

☎ Corsham (0249) 712214
4½ m SW of Chippenham on A4, turn S at Corsham
ST 8770 (OS 173)

Open mid Jan to mid Dec T, W, Th, S, Su 1400-1600; June to Sept and Bank Hols 1400-1800

🅿 WC ♿ (by appt) 🚻 (by appt) D (grounds only)
🍴 ⛱ ♨ (occasional) ✂ 𝄢 (by appt) ●

99

Lacock Abbey

near Chippenham, Wiltshire

Lacock owes its 18th-century appearance to the gentleman architect Sanderson Miller, who enlarged it in the 1750s in the 'Gothick' style, but in fact it has a much longer history. Originally an Augustinian convent, it was closed down in 1539 and bought by Sir William Sharrington, whose descendants have lived here ever since. Sharrington, unlike many of his contemporaries who built on old abbeys, retained a great deal of the original building, making most of his alterations upstairs where the living accommodation was (and still is). The result is that one enters a country house and finds oneself inside a medieval convent, with its lovely fan-vaulted cloister, stately chapter house and refectory. Among Sharrington's alterations that survived the 18th-century rebuilding are the octagonal tower, four storeys high, at the south-east corner of the building, and the stable block. The house is entered through Miller's splendid 'Gothick' entrance hall, and visitors can wander freely through the monastic buildings and the stable court as well as most of the first-floor rooms. Lacock is famous as the home of the pioneer of photography, William Henry Fox Talbot, who also built the oriel windows on the south front, the last alteration to be made to the house. Hanging by one of these windows is a photograph of it, a print from the negative made in 1835, and a converted barn houses the Fox Talbot Museum of the History of Photography. The Abbey and village were given to the National Trust by William Henry's granddaughter, Miss Matilda Fox Talbot, in 1944.

☎ Lacock (024 973) 227

4 m S of Chippenham on A350

ST 9168 (OS 173)

Open Apr to end Oct daily exc T and Good Fri 1400-1800; parties by appt at other times

♿ (exc Su) 🅿 WC 🚻 (by appt) ♣ 𝄞 NT Concerts

Littlecote House

near Hungerford, Wiltshire

Littlecote, a Tudor manor house set in an ancient park, is probably the oldest brick building in Wiltshire. In defiance of its name, it is a very large house, long and low, the entrance front of the 1590s being of the usual Tudor E-shape, but pulled out to a much greater width. The garden side, faced with flint, is older, dating from about 1490. Littlecote was owned by the Darrell family until 1589, when it was bought by Sir John Popham, later Lord Chief Justice of England, remaining in the Popham family until 1922. It was then acquired by Sir Ernest Salter Wills, whose grandson and family live here today. The Pophams were Puritans, and the great hall, a fine room with excellent panelling and furniture, displays Cromwellian weapons and uniforms, as well as a finger stock said to have been used by Sir John to confine prisoners in the dock. The chapel, one of the few examples of a Puritan religious building to survive complete, is typical in its arrangement, the elevated pulpit taking the place of the altar. The brick hall, so-called for its brick floor, has good 17th-century panelling, and there are some unusual paintings of nudes in Classical settings in the Dutch parlour. The bedrooms contain some fine oak furniture, including a four-poster bed contemporary with Elizabeth I's visit in 1601, and a marvellous set of crewel work bed curtains in the 'haunted bedroom', which is associated with a gruesome child murder by William Darrell in 1575. The lovely long gallery contains a good collection of furniture and family portraits.

☎ Hungerford (0488) 82170

2 m NW of Hungerford on B4192, continue on road to Littlecote

SU 3070 (OS 174)

Open Easter to end Oct daily 1000-1800

🅿 WC ♿ (limited access) ⊟ (by appt) D ♣ ♟
⋒ ◆ ⚶ 🏹 🐝 (not in house) 🚶

Longleat House

Warminster, Wiltshire

The first important Renaissance building in England, built by Sir John Thynne as a mansion fit for a visit by Elizabeth I, Longleat was designed with the help of Thynne's master mason, Robert Smythson. Sir John was able to entertain the Queen lavishly in 1575. The building is a vast square with two inner spaces designed as light-wells. The three-storeyed fronts have large mullioned windows, while the balustrades, chimneys and domed turrets above give a classical finish. Further additions to the exterior were made in the 17th century. Sir Thomas Thynne, who inherited in 1682, made many improvements to the house and to the gardens. One of the few that remains is Bishop Ken's library on the top floor, where most of the family's 30,000 books are housed. In the late 18th century Capability Brown was employed to landscape the park. The formal gardens were destroyed, lakes were formed from the stream known as the Long Leat, and the valley was planted with woods and clumps of trees. Britain's first safari park, complete with the famous lions, was opened here in 1966. The interior of the house has had several periods of alteration. The great hall is recognisably Elizabethan; the great staircase and lower east corridor were the work of Sir Jeffry Wyatville in the early 19th century, while the elaborate work in the state rooms and the red library was carried out in the 1870s. Longleat exhibits all the furnishings of a great house: notable silver and ceramics and exotic furniture.

☎ Maiden Bradley (098 53) 551

4 m W of Warminster on A362, turn S

ST 8043 (OS 183)

Open throughout year daily (exc 25th Dec) 1000-1800, closes 1600 Oct to Easter

🅿 WC ♿ (by appt) 🚻 D ♣ 🍴 ⅋ ◆ 🐾
♿ (also in French, German and Dutch) ⚡

Newhouse

Redlynch, near Salisbury, Wiltshire

Newhouse, originally called Tychebourne Park, was built for the owner of Longford Castle, Sir Edward Gorges, probably as a hunting lodge. It is a curious shape, in the form of a Y, the central part of which was completed by 1619 and the 'arms' extended in the 18th century. Only one other house of this shape is known. In 1633 Gorges sold the house to Giles Eyre, whose descendants still live there, and the two wings were added in 1742 and 1760. The furnishings have been collected over the years by the Eyre-Matcham family (Harriet Eyre married George Matcham in 1817) and are not exceptional, though there are some interesting pieces, including a child's cot made for Horatia Nelson, daughter of Lord Nelson and Lady Hamilton. Catherine Matcham was Nelson's sister, and more or less brought up Horatia. The parlour has Louis XVI armchairs with covers embroidered by Catherine Matcham, and the east room contains a lovely Queen Anne embroidered quilt. The house has some good family portraits, but the most interesting picture is the 'Hare picture', thought to have been painted about 1640. It apparently shows the triumph of hares over men, but it may have been a pun ('Hare' for 'Eyre') or an allegory of some kind. The drawing room was lavishly refurnished in 1906 in the 'Carolean' style by Maples, but the furnishings are now badly decayed and the room is being restored. It is used to display a collection of costume dating from the 1750s to the 1930s. One of the outhouses has collection of agricultural implements found nearby.

☎ Downton (0725) 20055

9 m S of Salisbury on A338, turn E onto B3080 then N to Redlynch

SU 2021 (OS 184)

Open Easter M and Spring Bank Hol M, May to Sept Su only and Bank Hol M 1400-1800

🅿 WC 🍴 (by appt) 🍽 (not in May) ✺ ●

Sheldon Manor

Chippenham, Wiltshire

This attractive stone manor house has a long history. The manor of Sheldon, as well as that of Chippenham, came to Sir Geoffrey Gascelyn in 1256, and his family held them until 1424, when Sheldon was bought by the Hungerfords, a powerful landowning family. They rebuilt one wing and added a chapel, but then evidently decided to live elsewhere, and let the house to the first in a long line of tenant farmers. By the end of the Civil War it was in a very run-down condition, and the current tenant, William Forster, had to obtain permission to rebuild it himself. In 1659 he rebuilt the entire left-hand side, with its tall gables and mullioned windows, but in 1684 the spendthrift Sir Edward Hungerford sold the property, and no further building was done, although two barns were added in the 1720s. Sheldon remained a farm until 1911, and the Gibbs family, the present owners, bought it in 1917 and have made it their home since 1957. The house has one of the finest 13th-century porches in the country, all that remains of the Gascelyn manor house. The first floor has a stone vault, and it contains the original stone water cistern, fed by pipes from the roof, while the room above, the priest's room, has a fine timber waggon roof. The rooms in the house are snug and lived-in, and there is a fine oak staircase in the 17th-century wing and the original tie-beams in the dining room. The interesting collection of furniture includes early oak pieces, Lancashire chairs, a country-made painted cupboard and two 17th-century Dutch cabinets.

☎ Chippenham (0249) 653120
1½ m NW of Chippenham on A420, turn S to Biddestone for 1 m
ST 8874 (OS 173)

Open Apr to Oct Th, Su, Bank Hols 1230-1800
🅿 WC ♿ (ground floor) 🍴 D ♣ ☕ ◆ 🐕 ✄
● (in house by permission) 🎗 (by request) ☂

Stourhead

Stourton, Warminster, Wiltshire

The house, built by Colen Campbell for the wealthy banker Sir Henry Hoare, and completed in 1725, is one of the earliest of the English Palladian mansions. Stourhead is famous for its grounds, which were laid out from 1744 by Henry Hoare II and the architect Henry Flitcroft, and take the form of an idealised Italian landscape; the house itself, lying on high ground above the lake, is also very fine. Colen Campbell's severely Classical stone house with its porticoed entrance forms the centre of the present building (the portico was actually added in 1840, but to a design left by Campbell). The main additions were made in the 1790s by Sir Richard Colt Hoare, grandson of Henry Hoare II. He was an antiquarian, scholar and amateur painter, and he built the two projecting wings, each containing one room, for his collection of books and paintings. The central block was gutted by fire in 1902, but well reconstructed. Sir Richard's wings, with their splendid Regency interiors, were unharmed. His library, Roman in inspiration, is a fine room with a barrel-vaulted ceiling and a carpet based on a mosaic pattern, and the furniture, including the flight of library steps, was designed and made by Thomas Chippendale the Younger, who made much of the furniture for the house. The picture gallery in the other wing, with a red and green colour scheme, has two tiers of windows on the east side, all with gilt and ebonised pelmets. Many of the paintings collected by the family remain, including Classical landscapes of the type that inspired the garden.

☎ Bourton (0747) 840348

14 m S of Frome on B3092, turn W at Stourton

ST 7734 (OS 183)

House open Apr and Oct S–W, May to end Sept daily exc F 1400-1730; gardens open daily throughout year 0800-1930 (dusk if earlier)

🅵 WC 🚻 (by appt) ♠ 🎪 ◆ ⚲ ● 𝘬 NT

Wardour Castle

Tisbury, Wiltshire

Wardour Castle, despite its name, is not a castle but an extremely fine Palladian house, built on the site of an older, once-fortified mansion. The Arundells, owners of the estate, were Catholics and improverished through the 17th and much of the 18th centuries, but in 1768 the 8th Lord Arundell married an heiress and was able to rebuild his ancestral home on a grand scale. The house, designed by James Paine and built between 1770 and 1776, is most imposing, with a massive central block containing splendid reception rooms and the two wings containing a Catholic chapel on one side and kitchens on the other. There is an impressive central pediment on the garden side. The low, columned entrance hall leads directly into the marvellous 'Pantheon' or staircase hall, the best feature of the interior, where the two semi-circular flights of the staircase curve up to a first-floor gallery, and the dome above is carried by eight soaring columns. The house is now a girls' school, and the contents and many of the mural decorations have been dispersed, but there are some family portraits, and the finely proportioned rooms contain fine plaster decoration, the most notable example being the ceiling of the music room. The chapel, part of Paine's original design but extended by Sir John Soane in 1788, is rich and splendid. The decorator was an Italian called Quarenghi, who may also have been responsible for the music room ceiling. The romantic ruins of Wardour Old Castle can be seen from the house.

☎ Tisbury (0747) 870464

21 m SW of Salisbury on A30, turn N to Donhead St Andrew

ST 9227 (OS 184)

Open 27 July to end Aug M, W, F, S 1430-1700; telephone before visit

🄿 WC 🚻 (by appt; 50 max) ♿

🕴 (compulsory) ●

Wilton House

Wilton, Wiltshire

Three miles west of Salisbury, Wilton House stands on the site of a Saxon abbey. In 1544 Henry VIII gave the abbey and lands to William Herbert, who had married the sister of Catherine Parr, the King's sixth wife. Herbert, later 1st Earl of Pembroke, pulled down the abbey and built a new house. In about 1647 a fire destroyed most of the Tudor house and its contents. Rebuilding work began at once. Inigo Jones had already done some work at Wilton, and the new house was constructed under his supervision. It was completed in about 1653. The east and west fronts date from this time and have the restrained dignity characteristic of Jones' work. The north and west blocks were, however, remodelled in Gothick style at the beginning of the 19th century by James Wyatt. The grandest room at Wilton, the 'double cube', is regarded as the best-proportioned room in England. It measures 60 foot by 30 foot, and is 30 foot high. The room and its setting provide a superb setting for the family portraits painted by Van Dyck for the 4th Earl. The ceiling has a flamboyantly painted coving; there are gilded friezes, exotic wall decorations, gilt and red velvet furniture by Chippendale and William Kent. There are noble rooms – the great ante-room, the white and gold colonnade room and the corner room all display outstanding paintings, by Andrea de Sarto, Reynolds, Rembrandt, Rubens, Poussin, as well as other treasures such as fine Chinese porcelain and a 250-piece Viennese dinner service.

☎ Salisbury (0722) 743115

3 m W of Salisbury on A36

SU 0931 (OS 184)

Open Apr to mid Oct T-S and Bank Hol M 1100-1800, Su 1300-1800

⊖ P WC ☒ ☐ (by appt) D (guide dogs only) ♠
� ◆ 🎴 ⚗ ✗ (compulsory exc Su) ● ⵝ

Avon

Blaise Castle House, Henbury, Bristol, Avon (tel Bristol [0272] 506789). 4 m NW of Bristol on A4018. ST 5678 (OS 172). 18th-century house containing social history museum covering West Country life from 1750 to 1900. Extensive picturesque gardens laid out by Humphry Repton with buildings by John Nash. Blaise Hamlet, also by Nash, is nearby. Open S-W am and pm all year, exc Christmas Day, Boxing Day, New Years Day and Good Friday. ⊖ 🅿 WC ♿ ⊟ (by appt) ♣ ☕ ⛩ ★ ◆ ✲ 🚶 🚶

Claverton Manor, near Bath, Avon (tel Bath [0225] 60503). 3¾ m SE of Bath on A36, turn W. ST 7864 (OS 172). Classical villa built c. 1820 and now housing the American Museum in Britain, with exhibits illustrating many aspects of American domestic life from 17th to 19th centuries, notably of American Indians, Pennsylvania Germans and Shakers. Open daily (exc M) pm, am and pm Bank Hols and preceding Su, end Mar to end Oct. Other times by appt. ⊖ 🅿 WC ♿ (limited access) ⊟ (by appt) D (on lead in grounds) ♣ ☕ ◆ ✲ 🚶 ●

Clevedon Court, Clevedon, Avon (tel Clevedon [0272] 872257). 1 m E of Clevedon on B3130. ST 4271 (OS 172). Manor house originally built in the 14th century, with a restored hall and 14th-century chapel. The house was extended continuously over the centuries. It has associations with Charles Makepeace Thackeray, and with Sir Edmund Elton, a well-known late 19th-century potter. Open Apr to end Sept W, Th, Su and Bank Hol M pm. ⊖ 🅿 WC ♣ ✲ 🚶 ● NT

Georgian House, 7 Great George St, Bristol, Avon (tel Bristol [0272] 299771 ext 237). Close to city centre. ST 5972 (OS 172). Late 18th-century town house now completely restored to recreate the home of the wealthy merchant and West Indies sugar plantation manager, John Pinney. The kitchen is particularly interesting. Open daily (exc Su) 1000-1300 and 1400-1700. ⊖ ♿ ⊟ (by appt) ★ ✲ 🚶 (by appt)

Horton Court, Horton, Avon. 3 m NE of Chipping Sodbury on A46, turn W. ST 7684 (OS 172). Cotswold manor house, with north wing and hall built c. 1140. There is a 16th-century porch, and an ambulatory or covered walk in garden. Open W and S 1400-1800, Apr to end Oct, other times by appt. 🅿 ♿ ⊟ ♣ ✲ ● (no flash) NT

Red Lodge, Park Row, Bristol, Avon (tel Bristol [0272] 299771). Close to city centre. ST 5972 (OS 172). House built c. 1570 by merchant John Yonge; upper parts rebuilt in 18th century. The interiors are the oldest in Bristol; they contain original panelling and plasterwork, and 17th- and 18th-century furniture. Open M-S am and pm. ⊖ ⊟ (by appt) ★ ♣ (certain times only) ✲

1 Royal Crescent, Bath, Avon (tel Bath [0225] 28126). Close to city centre. ST 7465 (OS 172). Stone-built Georgian house in the Crescent built by John Wood the Younger in 1767. Interior carefully restored in 1970 and furnished in authentic 18th-century style. Open daily (exc M) am and pm March to Dec; (S pm only). ⊖ WC ⊟ (by appt) ◆ ✲ 🚶 ●

Cornwall

Godolphin House, near Helston, Cornwall (tel Germoe [073 676] 2409). 2 m NW of Helston on A394, turn N onto B3302 for 3 m then W. SW 6031 (OS 203). 16th- and 17th-century house built for the Earls of Godolphin, with unusual granite colonnaded front of 1635. The house is said to have been used by the future King Charles II on his escape from Pendennis Castle. Open May and June Th pm; July and Sept T and Th pm, Aug T pm, Th am and pm. 🚹WC🚻 (by appt) ♣ ✗ (for parties)

Lanhydrock, near Bodmin, Cornwall (tel Bodmin [0208] 3320). 2½ miles SE of Bodmin. SX 0863 (OS 200). House built in 1630-42, and rebuilt 1881 after a fire. The gatehouse and finely decorated long gallery are 17th-century; the rest of the interior has Victorian decoration. Extensive kitchens. The 19th-century formal gardens lead down to the River Fowey. House and gardens open 1st Apr to 31st Oct daily 1100-1800; Nov to March, gardens only. ⊖🚹WC🚻 (by appt) D ♣ 💺 NT

Lawrence House, 9 Castle St, Launceston, Cornwall (tel Launceston [0566] 2833). Close to town centre. SX 3284 (OS 201). 18th-century house, now containing a museum of curious items of local interest, including the feudal dues owed by the town to the Duke of Cornwall. Open Apr to Sept M-F (not Bank Hols) 1030-1230, 1430-1630. ⊖ WC 🚻 ● (no flash) NT

Mount Edgcumbe House, near Plymouth, Cornwall (tel Plymouth [0752] 822236). By pedestrian ferry from Plymouth, or on B3247. SX 4552 (OS 201). 16th-century house adapted in the 18th century and rebuilt after the Second World War. The large 18th-century grounds are varied, with formal gardens, open parkland and mature woods, and enjoy outstanding views over Plymouth Sound. Enquire for details of house opening; gardens daily all year. ⊖ WC 🦽 🚻 D ♣ 💺 ⊞ ★ ◆ ⚘ ✗ (by appt) 🚶

Old Post Office, Tintagel, Cornwall. In village centre, 13 miles N of Bodmin, on B3263. SX 0588 (OS 200). Medieval 14th-century manor house, with large hall and parlour. It was used between 1844 and 1892 as a village post office, and has now been restored as such. Open 1st Apr to 31st Oct daily 1100-1800. ⊖🚹🦽 NT

Trelowarren House, Mawgan in Meneage, near Helston, Cornwall (tel Mawgan [032 622] 366). 3 m SE of Helston on A3083, turn NE onto B3293 for 2 m then NE. SW 7125 (OS 203). Home of the Vyvyan family since 1427, the main part of the house dates from the early 16th century. The notable chapel was rebuilt in the 18th century in a bright Gothick style. Part of the house is used by the Trelowarren Fellowship, an Ecumenical Charity. House open Easter to Oct, W and Bank Hols pm; also last Su July to 1st Su Sept pm. Chapel open daily am and pm. 🚹WC🦽 (limited access) 🚻 (by appt) ♣ 💺 ⊞ ◆ ⚘ ✗ (compulsory)

Trerice, St Newlyn East, near Newquay, Cornwall (Newquay [063 73] 5404). 4½ m SE of Newquay on A3058, turn SW. SW 8458 (OS 200). Manor house built by Sir John Arundell in 1572 on an E-plan. The interior contains a collection of 17th- and 18th-century furniture. The barn contains an exhibition tracing the history of the lawn mower; the garden has only recently been planted. Open 1st Apr to 31st Oct daily 1100-1800. ⊖ 🚹 WC 🦽 🚻 (by appt) ♣ 💺 NT

Trewithen, near Probus, Cornwall (tel St Austell [0726] 882418). 6 m E of Truro on A390. SW 9147 (OS 204). Early 18th-century house, built by Thomas Edwards and decorated in Rococo style. The gardens were laid out by George Johnstone in the 1920s and are known internationally for their camellias, rhododendrons, magnolias and other exotic trees and shrubs. House open Apr to July M and T pm; garden open March to Sept daily (exc Su) pm. ⊖ 🚹WC 🦽 🚻 (by appt) ♣ 🐝 ⚘ (garden only) ✗ (house only) ● (not in house)

Devon

Arlington Court, Barnstaple, Devon (tel Shirwell [027 182] 296). 8 m NE of Barnstaple on A39, turn E. SS 6140 (OS 180). Neo-Classical style Regency house virtually unchanged since the mid-19th century and containing the collection of its last private owner, Miss Rosalie Chichester, which includes shells, pewter, model ships, objets d'art and costume. The stables include a collection of 19th-century vehicles. There is a formal garden and a park. House open Apr to Oct daily (exc S) 1100-1800; garden and park open all year 1100-1800 (dusk if earlier). ⊖ 🅿 WC 🔣 🛏 (by appt) ♣ ☞ ⚭ NT

Bowden House, Totnes, Devon (tel Totnes [0803] 863664). 1 m S of town centre on A381, turn SE. SX 8058 (OS 202). Ancient house, with parts dating back to the 13th century but primarily an early 18th-century façade on an early Tudor mansion. The hall is decorated in Baroque style. Open Apr to Oct T, W and Th pm; all year for parties by appt. ⊖ 🅿 WC 🛏 (by appt) ♣ ☞ ◆

Bradley Manor, Newton Abbot, Devon. On west side of Newton Abbot, on A381. SX 8470 (OS 202). Small manor house mainly dating from about 1420. The interior contains many architectural features of the 15th and 16th century, including great hall, buttery and chapel. Open Apr to Sept W 1400-1700, also Th Apr 2 & 9, Sept 17 & 24 (1987) 1400-1700. ⊖ 🅿 🔣 (limited access) 🛏 (by appt) ♣ ⚭ ● (not in house) NT

Buckland Abbey, near Yelverton, Devon (tel Yelverton [0822] 3607). 7 m SE of Tavistock on A386, turn SW. SX 4866 (OS 201). 13th-century Cistercian monastery, bought by Richard Grenville in 1541, converted in the 1570s, and bought by Sir Francis Drake in 1581 after his voyage around the world. The house now contains relics of Drake and the Grenville family, including 'Drake's Drum', naval exhibits and a Devon folk museum. The grounds contain a large tithe barn. Open Easter to end Sept daily 1100-1800 (then closed until July 1988) ⊖ 🅿 WC 🔣 (limited access) 🛏 (by appt; phone Plymouth Museum [0752] 668000) D (grounds only) ♣ ☞ ◆ ● (in house with permission only) NT

Cadhay, Ottery St Mary, Devon (Ottery St Mary [040 481] 2432). 5 m SW of Honiton on A30, turn SE. ST 0896 (OS 192). Mid-Tudor house, built from proceeds of the Dissolution of the Monasteries; extended around 1600, and with a new front added in the 1730s. Open late Spring and Summer Bank Hols, and July to Aug, T-Th pm. ⊖ (1 mile) 🅿 WC 🛏 (by appt on other days) D (on lead in grounds) ♣ ⚮ 🍴

Castle Hill, Filleigh, Barnstaple, Devon (tel Filleigh [059 86] 227). 3½ m W of South Molton on A361. SS 6728 (OS 180). Grand Neo-Classical house built 1730-40, family home of the Fortescue family. The house contains a notable collection of 18th-century furniture and *objets d'art*. The gardens are extensive and varied, and include an aboretum. Open Apr to Oct by appt. ⊖ 🅿 WC 🛏 D (on lead in grounds) ♣ ☞ (by appt) ⚲ 🍴 (compulsory) ●

Chambercombe Manor, Ilfracombe, Devon (tel Ilfracombe [0271] 62624). 1 m SE of Ilfracombe on A399, turn S. SS 5346 (OS 180). Small and attractive early medieval manor, mainly rebuilt in the 16th and 17th centuries, and containing furniture of that period. Open Easter to end Sept M-F am and pm, Su pm only. ⊖ (½ mile) 🅿 WC 🔣 (limited access) 🛏 (by appt) ♣ ☞ ◆ ⚮ 🍴

Dartington Hall, near Totnes, Devon (tel Totnes [0803] 862271). 2 m NW of Totnes, on A384. SX 7962

(OS 202). Late 14th-century house, with original great hall, outer courtyard and other buildings. Since the 1920s the house has been devoted to an educational trust, providing a progressive school and many craft activities, notably glassware. Gardens only open daily throughout year (also Great Hall if free). ⊖ ▣ wc ⊟ (by appt) ♠ ☞ ★ ◆ (in village) ⚔ (by appt; tel [0803] 863614)

Elizabethan House, 32 New St, Plymouth, Devon (tel Plymouth [0752] 668000 ext 4380). In town centre, off the quay. SX 4854 (OS 201). Elizabethan merchant's house, restored in 1926. Its staircase is built around a ship's mast. The house contains 16th-century furniture. Open June to Sept daily am and pm (Su pm only); Oct to Apr closed Su. ⊖ wc ⊟ ♠ ◆ ⚹

Flete, Ermington, Ivybridge, Devon (tel Holbeton [075 530] 308. 11 m E of Plymouth on A379, turn S. SX 6251 (OS 202). Jacobean house rebuilt by Richard Norman Shaw in the 19th century. Open May to Sept W and Th 1400-1700. ⊖ ▣ wc ⊟ (by appt) ♠ ⚹ ⚔

Killerton House, near Exeter, Devon (tel Exeter [0392] 881345). 6 m NE of Exeter on A396, turn E onto B3185. SS 9700 (OS 192). Georgian house erected in the late 1770s for the Acland family. The interior contains an extensive collection of period costume. The grounds were laid out around 1800 by John Veitch with many rare trees and sweeping lawns. House open Apr to end Oct daily am and pm; grounds open all year daily am and pm. ⊖ (¾ m)▣ wc ▣ ⊟ (by appt) D (park only) ♠ ☞ ◆ ⚹ NT

Kirkham House, Kirkham St, Paignton, Devon (tel Paignton [0803] 522775). In town centre. SX 8860 (OS 202). Late medieval manor house, now restored and known as the Priest's House. It contains a collection of locally made furniture. Open Apr to 1st Oct daily exc T am and pm (Su and W pm only). ⊖ ▣ ⊟ D ♠ ⚹ EH

Knightshayes Court, near Tiverton, Devon (tel Tiverton [0884] 254665). 3 m N of Tiverton on A396. SS 9615 (OS 181). Victorian house begun by William Burges in 1869 in Gothic Revival style, and decorated by J.D. Crace. The gardens are highly regarded. Gardens open Apr to end Oct daily exc F am and pm; house pm only. ▣ wc ▣ ⊟ (by appt) ♠ ☞ ⼌ ◆ ⊛ ⚹ NT

Tapeley Park House, Instow, Devon (tel Instow [0271] 860528). 1 m S of Instow, on A39. SS 4729 (OS 180). House built in the early 18th century, and modified around 1900. There are Georgian interiors, a collection of porcelain and distinctive plasterwork. The beautiful terraced Italian gardens were laid out in the early 20th century, and the house overlooks the estuary of the Rivers Taw and Torridge. Open Apr to Oct daily (exc M) am and pm; Nov to Apr gardens only open. ⊖ ▣ wc ▣ ⊟ (by appt) D ♠ ☞ (parties only) ⼌ ⊛ ⚔ ●

Torre Abbey, Torquay, Devon (tel Torquay [0803] 23593). On Torquay sea front. SX 9063 (OS 202). 12th-century monastery converted into a private house after the Dissolution of the Monasteries, and much rebuilt in the 18th century. It contains period rooms and a large collection of paintings and other works of art. The medieval ruins and tithe barn can also be visited. Open Apr to Oct daily am and pm. ⊖ wc ⊟ ♠ ⼌ ⚔ (by appt)

Ugbrooke, near Chudleigh, Devon (tel Chudleigh [0626] 852179). 12 m SW of Exeter on A38, turn S onto A380 for 7 m then NW. SX 8778 (OS 192). Medieval house substantially rebuilt in the 1770s by the young Robert Adam in his 'castle' style for the Clifford family. Collection of embroidery and tapestry. There is a separate private chapel. The grounds were laid out by Capability Brown. Open Spring Bank Hol to Aug Bank Hol S, Su and Bank Hols only pm. ⊖ (1 mile) ▯ WC ⬙ ⊟ D (on lead) ♠ ☞ ⋔ ⋇ ⨍ (compulsory, at set times) ●

Dorset

Chettle House, Blandford Forum, Dorset (tel Tarrant Hinton [025 889] 209). 8 m NE of Blandford Forum on A354, turn NW. ST 9513 (OS 195). House in the English Baroque style, built by Thomas Archer in the 1710s. Much of the house has been modified, but the staircase remains a dramatic example of the style. Open daily exc T 1030-1730. ▯ WC ⊟ (by appt) ♠ ⊛ ⋇ ⨍ (by appt)

Cloud's Hill, near Wool, Dorset. 9 m E of Dorchester. SY 8290 (OS 194). Cottage that was the home of T.E. Lawrence (Lawrence of Arabia) from 1925 when he rejoined the RAF under an assumed name. The house contains a collection of items relating to Lawrence. Open Apr to Sept W-F, Su and Bank Hol M 1400-1700; Oct to March, Su 1300-1700 only. ▯ NT

Compton House, near Sherborne, Dorset (tel Yeovil [0935] 74608). 4 m W of Sherborne on A30. ST 5916 (OS 183). 16th-century manor house greatly embellished in period style in the 19th century, now used to breed and display butterflies. These can been seen in all stages of development and several different environments. There is also a silk farm at which silk worms and silk-making can be seen. Open Apr to end Oct daily am and pm. ⊖ ▯ WC ♠ ☞ ⋔ ◆ ⋇

Deans Court, Wimborne, Dorset. Close to town centre. SZ 0199 (OS 112

195). House built as the deanery for the Minster, and extended in the 18th century. The gardens are partly wild, and retain the monastery fishpond and an 18th-century kitchen garden. There are interesting trees and a sanctuary for threatened vegetables. House open by appt only; gardens open Easter, Spring and Summer Bank Hols Su (pm) and M (am and pm); last Su June, July, Sept (pm). ⊖ ▯ (for disabled) WC ⬙ ⊟ ♠ ☞ ◆ ⋇ ⨍ (by appt)

Hardy's Cottage, Higher Bockhampton, Dorset (Dorchester [0305] 62366). 4 m E of Dorchester on A35, turn S. SY 7292 (OS 194). Small thatched cottage in which the novelist and poet Thomas Hardy was born in 1840. The interior is little altered since that time. Open by appt only, Apr to Oct am and pm. ⊖ ▯ ⊟ ⋇ ● (no flash) NT

Milton Abbey, near Blandford Forum, Dorset (tel Milton Abbas [0258] 880258). 12 m SW of Blandford Forum on A354, turn N. ST 7902 (OS 194). 14th-century abbey church and, beside it, a Georgian Gothick house constructed by William Chambers and James Wyatt from the original hall (completed 1498) of the abbey. The house now is used as a school, for which the hall is the dining room. Its carved screen and hammer-beam roof are notable. House open school hols: 2 weeks at Easter, late July to end Aug daily 1000-1830; church open daily throughout year 1000-dusk. ▯ WC ⊟ (by appt) D (on lead) ♠ ☞ ⋇

Purse Caundle Manor, Purse Caundle, Dorset (tel Milborne Port [0963] 250400). 5 m E of Sherborne on A30, turn SE. ST 6917 (OS 183). Typical manor house, mainly of the mid-16th century but with a somewhat earlier hall, and with interesting woodwork throughout. Open Easter to end Sept Th, Su and Bank

Hols pm, other times by appt. ⊖ 🅟
🅗 (by appt) ♣ 🗣 (by appt) ⁂
𝙆 (compulsory) ●

Sandford Orcas Manor House,
Sandford Orcas, Dorset (tel Corton
Denham [096322] 206). 5 m E of
Yeovil on A30, turn N at Sherborne.
ST 6221 (OS 183). Fine Tudor manor
house of stone, preserved as it was
in about 1500. The interior contains
a fine collection of stained glass and
16th-century furnishings. Open
Easter M and May to Sept M am and
pm, Su pm only. 🅟 WC 🅗 (by appt)
♣ ⁂ ● (not in house)

Smedmore House, Kimmeridge,
Dorset (tel Corfe Castle [0929]
480717). 6 m S of Wareham on A351,
turn SW.SY 9278 (OS 195). Manor
house begun in the 1620s (for a
project involving the use of the local
oil shale for a glass factory), but
mainly of the Queen Anne and
Georgian period. It contains Dutch
marquetry furniture and paintings,
and a collection of antique dolls.
There is a walled garden and
interesting plants and shrubs. Open
1st W in June to 2nd W in Sept and
Aug Bank Hol Su 1415-1730. 🅟 WC
🅗🅗 (by appt, preferably Th) D (on
lead, grounds only) ♣ 🗣 (limited)
𝙆 (by appt) ⁂ ● (not in house)

Somerset

Barford Park, near Enmore,
Somerset (tel Spaxton [027 867] 269).
5 m N of Bridgwater on A39, turn
SW then N. ST 2335 (OS 182).
Miniature Queen Anne period red-
brick country house in a park with
fine trees. There is a formal garden
and a water garden. Open early
May to end Sept F-Su and Bank Hol
M 1400-1800. 🅟 WC 🅗 (by appt) ♣ ●

Barrington Court, Ilminster,
Somerset (tel Ilminster [046 05]
41480). 3 m NE of Ilminster on
B3168, turn SE. ST 3918 (OS 193).
Attractive 16th-century E-shaped
house, restored in the early 20th

century. The gardens were laid out
by Gertrude Jekyll in the 1920s, and
there is a late 17th-century stable
block. House open mid Apr to end
Sept W, Su 1415-1700; gardens open
mid Apr to end Sept Su-W 1400-
1700. ⊖ 🅟 WC 🅗 🅗 (by appt) ♣ 🗣 ⌂
◆ ❀ ⁂ 𝙆 (by appt) ● ⚡ NT

Bishop's Palace, Wells, Somerset
(tel Wells [0749] 78691). In city
centre. ST 5545 (OS 182). Fortified
and moated palace, dating in part to
the 13th century, and with a 14th-
century gatehouse. The great hall,
now ruined, was built in 1280.
The interior of the palace was
remodelled in the 17th century,
with a fine staircase and the creation
of a long gallery; it was restored in
the mid-19th century. Open Easter
to Oct Th, Su and Bank Hol M pm;
Aug daily pm. ⊖ WC 🅗 (ground floor
only) 🅗 🗣 ◆ ⁂ 𝙆

Coleridge Cottage, Nether Stowey,
Somerset (tel Nether Stowey [0278]
732662). 8 m W of Bridgwater, on
A39. ST 1939 (OS 181). The home of
the poet Samuel Taylor Coleridge
from 1797 to 1800, during which he
wrote *The Ancient Mariner*. The
parlour and reading room are open
to the public. Open Apr to end Sept
T-Th, Su pm. ⊖ 🅟 🅗 𝙆 NT

King John's Hunting Lodge,
Axbridge, Somerset (tel Axbridge
[0934] 732012). 12½ m NW of Wells
on A371. ST 4255 (OS 182). Early
15th-century merchant's timber-
framed house. It has no apparent
connection with King John, or with
hunting, but it was used as an
alehouse in the 17th and 18th
centuries. The house is now a
museum of local history. Open Apr
to Sept daily 1400-1700. ⊖ 🅟 🅗 (by
appt) ★ ⁂ 𝙆 (by appt) ● (no
flash) NT

Lytes Cary, Ilchester, Somerset (tel
Castle Cary [04582] 23297). 2½ m NE
of Ilchester off A372. ST 5326 (OS
183). Manor house built in the 14th

to 16th centuries, the home of Henry Lyte, the Elizabethan herbalist. The chapel is the oldest part of the house, and the hall is mid-15th century, with a timber roof. The house is furnished in styles ranging from the 16th to 18th centuries, and the gardens have been replanted in the fashion of about 1600. Open Apr to end Oct W and S 1400-1800. ⊖ (1½ miles) P ♿ ⊟ (by appt) ♣ ⁂ ● (no flash) NT

Midelney Manor, Drayton, near Langport, Somerset (tel Langport [0458] 251229). 15 m E of Taunton, on A378. ST 4024 (OS 193). Medieval manor belonging to the abbot of nearby Muchelney Abbey, which became a private house in the 1530s. The house dates from the 1540s. There is an unusual 17th-century falconry mews, and a heronry in the gardens. Open June to Sept W 1400-1730; Bank Hol M 1400-1730. ⊖ P WC ♿ (limited access) ⊟ (by appt) ♣ ☕ (parties only, by appt) ⁂ 𝒻

Oakhill Manor, Oakhill, Somerset (tel Oakhill [0749] 840210). 4 m N of Shepton Mallet on A37, turn E. ST 6347 (OS 183). Attractive country house in 45-acre park, containing a notable collection of model transport exhibits, covering air, land and sea transport. Miniature railway through gardens. Open daily Easter to Oct pm. P WC ⊟ ♣ ☕ ⛩ ◆

Priest's House, Muchelney, Somerset (tel Langport [0458] 250672). 17 m E of Taunton on A378, turn SE. ST 4325 (OS 193). Late medieval house with large Gothic hall window, originally the residence of the parish priest. Open by appt only. ⊖ (1 mile) NT

Tintinhull House, Tintinhull, near Yeovil, Somerset (tel Martock [093 582] 2509). 5 m NW of Yeovil on A303, turn S. ST 5019 (OS 183). House of 1600 with an elegant façade of the 1720s. Its gardens were laid out by the botanist Dr Price in the early 20th century. Open Apr to end Sept W, Th, S and Bank Hol M pm. P WC ♿ ⊟ (by appt) ♣ ⁂ NT

Wiltshire

Avebury Manor, Avebury, Wiltshire (tel Avebury [067 23] 203). 15 m SW of Swindon on A4361. SU 0970 (OS 173). 16th-century manor house built on the site of an old monastery. The interiors are 16th-century to Victorian and contain fine plasterwork and fireplaces. The woodwork of the dining room is notable, and the oak staircase is the work of the 17th century. The gardens are laid out in formal Elizabethan style. Open Apr to end Sept daily pm 1130-1800 M-S, 1230-1730 Su; Oct to end March S and Su only; other times by appt. ⊖ P WC ♿ (limited) ⊟ ♣ ☕ ⛩ ◆ ⁂ 𝒻 (parties only) ●

Bowood House, Calne, Wiltshire (tel Calne [0249] 812102). 5 m SE of Chippenham on A4, turn S. ST 9769 (OS 173). 18th-century house designed by Robert Adam, once the 'Little House' to the main Bowood House which was demolished in 1955. It houses Adam's library, and the room in which Joseph Priestley discovered oxygen. The old orangery now houses a picture gallery, with Old Master works; there is also a collection of Indian silver caskets, and antique sculpture. The beautiful grounds were laid out by Capability Brown and Humphry Repton in the 1760s; among the features are a rose garden, an Italian garden, an 18th-century aboretum, a mausoleum by Nash, a temple, and a lake. Open Apr to Sept daily am and pm. ⊖ (1½ miles) P WC ♿ ⊟ ♣ ☕ ⛩ ◆ ❀ ⁂ 🏃

Chalcot House, Westbury, Wiltshire (tel Chapmanslade [037 388] 466). 2 m W of Westbury, on A3098. ST 8448 (OS 183). Small late 17th-century Classical manor house, recently restored and with attractive

18th- and 19th-century furniture. Open Aug daily 1400-1700. P WC ⊟ (by appt) ♣ ☕ (by appt) ⊼ ⁂ ✗

Great Chalfield Manor, Melksham, Wiltshire. 2½ m NE of Bradford-on-Avon off B3109. ST 8563 (OS 173). Manor house of the late 15th century with a moat, adjoining a parish church of the same date. There is a great hall, with a restored screen and a contemporary mural of the builder. Open Apr to Oct T-Th 1200-1300 and 1400-1700. ⊖ (1½ m) P ⊟ (by appt) ♣ ⁂ ✗ ● (no flash) NT

Lydiard Park, Purton, Wiltshire (tel Purton [0793] 770401). 4 m W of Swindon on A420, turn N. SU 1084 (OS 173). Mid-Georgian Classical house, with fine plasterwork interiors and contemporary furniture. The parish church is set in the extensive park. Open daily am and pm (Su pm only). ⊖ (1 mile) P WC ♿ ⊟ ♣ ☕ ⊼ ⁂ ✗ (by appt) ✱ ✱

Mompesson House, Choristers' Green, Cathedral Close, Salisbury, Wiltshire (tel Salisbury [0722] 335659). In city centre. SU 1429 (OS 184). Small stone house built in 1701, a fine example of Georgian provincial town architecture. The panelling and plasterwork are impressive; the oak staircase was added in the 1740s. There is also a collection of English 18th-century drinking glasses, and an attractive garden. Open Apr to end Oct M-W, S and Su 1230-1800. ⊖ ♿ ⊟ ♣ ⁂ ● (no flash) NT

Philipps House, Dinton Park, Dinton, Wiltshire (tel Teffont [072 276] 208). 12 m W of Salisbury on B3089. SU 0031 (OS 184). Neo-Classical house built in the early 19th century by Jeffry Wyattville for the Wyndham family. The house is now used as a conference centre for the YWCA. Open Apr to Sept by appt only. ⊖ P NT

Pythouse, Tisbury, Wiltshire (tel Tisbury [0747] 870210). 4½ m N of Shaftesbury, 2½ m W of Tisbury. ST 9028 (OS 184). Palladian-style early 19th-century mansion. The reception rooms contain large 16th-century Italian fireplaces. There is an extensive garden with an orangery. Open May to Sept W and Th 1400-1700 (last admission 1600). ⊖ (limited) P WC ⊟ (by appt) ♣ ⁂ ● (not in house)

Westwood Manor, Bradford-on-Avon, Wiltshire (tel Bradford-on-Avon [02216] 3374). 1½ m SW of Bradford-on-Avon on B3109, turn W. ST 8159 (OS 173). 15th-century stone manor house, with Tudor and Stuart additions. The 'king's room' and great chamber contain excellent plasterwork of the early 17th century. There is also a modern topiary garden. Open Apr to end Sept Su, M 1400-1730; specialist tours 1100 and 1430, 1st W in each month by appt). ⊖ (1½ miles) P ⊟ (by appt) ♣ ⁂ ● NT

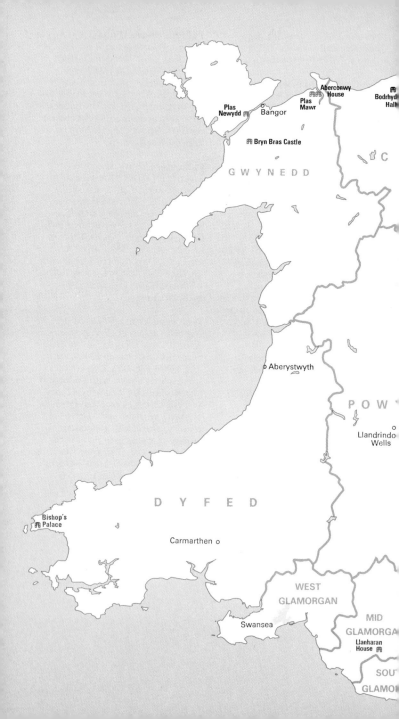

Wales and Western Counties

Adlington Hall

Macclesfield, Cheshire

This small manor house, an attractive mixture of brick and the black-and-white timbering so common in Cheshire, has been the home of the Legh family since 1315, and additions and alterations have been carried out by successive generations. The timber parts of the building date mainly from the 16th century, and the north front seems to have been rebuilt in the mid-17th century. Extensive alterations were made after 1739 by Charles Legh, who added a west wing, several rooms in the north-west corner, and a Georgian south front connecting the new wing with the old Elizabethan east wing. In 1928 part of the west side was removed, the house having been found unmanageably large. Visitors see the Elizabethan front first, but enter through a door in the north front into a screens passage leading to the Elizabethan staircase. After passing through some charming small panelled rooms on the first floor of the north range, we see the attractive 18th-century drawing room and dining room, and then the lofty and impressive great hall, the centre from which the house grew. Completed in 1505, the hall was given new and bigger windows in 1581, and has a fine hammerbeam roof. Its most striking features, however, are the twin oak trees supporting the east end, all that remains of the original hunting lodge. These have been adze-carved into octagonal shape, and between them is one of the finest 17th-century organs in the country, which was built by 'Father' Bernard Smith in about 1670 and is richly ornamented.

☎ Macclesfield (0625) 829206

9 m S of Stockport on A6 turn S on A523

SJ 9080 (OS 109)

Open Good Fri to end Sept Su and Bank Hols; Aug W, S, Su and Bank Hol M 1400-1730

⊖ (1 m) 🅿 WC ♿ 🚻 (by appt at other times) D (on lead) ♥ 🍽 🎏 ◆ ⚘ 𝄢 (parties by appt)

Bramall Hall

Bramhall Park, Bramhall, Stockport, Cheshire

This unusually large timber-framed house of the black-and-white Cheshire type was built mainly between 1500 and 1600 for the Davenport family. The buildings were originally arranged round a central courtyard, but one entire side was removed in the 18th century. The Industrial Revolution in the 19th century brought considerable wealth to the area through the cotton, steel and hatting industries, and a newly-rich Victorian owner, Charles Neville, carried out extensive repair and restoration work on the house. There are some extremely fine rooms, the ballroom in the south wing being particularly notable for its Tudor and Stuart wallpaintings depicting 16th- and 17th-century life. The withdrawing room has a lovely Tudor plaster ceiling and fireplace, and the walls are hung with family portraits. 'Dame Dorothy's bedroom', with its plaster frieze and secret 'hide', dates from the same period. The great hall has early stained glass, and there is more in the chapel, which also contains wallpaintings on the theme of the Reformation. The house, which is administered by Stockport Borough Council, is not lived in, and there is not a great deal of furniture, but there are some good early pieces, and the 'Chapel Room' contains some excellent furniture by A. W. N. Pugin. The park is landscaped, and extends to some 60 acres.

☎ (061) 485 3708

3 m S of Stockport on A5102 at Bramhall Park

SJ 8986 (OS 109)

Open daily exc M (but inc Bank Hol M) Apr to Sept 1200-1700; Oct to Mar (closed Jan) 1200-1600

WC

Gawsworth Hall

Macclesfield, Cheshire

A pretty half-timbered manor house with a fine three-storey bow window in its long show front. The Hall dates mainly from the 15th century, though additions were made later, and the south and east walls were rebuilt in brick in the 18th century. The rooms are relatively small and many contain fine furniture, paintings and sculpture. The library, a double cube room measuring 16 by 32 feet, has a lovely carved Tudor chimneypiece, and the bookcases were designed by A. W. N. Pugin. The chapel contains some stained glass by William Morris and Burne-Jones. The outbuildings house a collection of 19th-century vehicles, the centrepiece being three Victorian horse-drawn double-decker buses. The park, encompassed by its Tudor wall, remains much the same as in medieval times, when it was the scene of knightly tournaments. Gawsworth Hall, now the home of the Roper-Richards family, has seen many stirring events since the long tenure (1316-1662) of its original owners, the 'fighting Fittons'. Mary Fitton, the 'wayward maid of Gawsworth', and thought to have been Shakespeare's 'Dark Lady', lived here, and in 1701 a famous duel was fought here between Lord Mohun and the Duke of Hamilton. Both were killed. The last professional jester in England, 'Magotty Johnson', was dancing master to the children of the house, and lies buried in a nearby spinney.

☎ North Rode (026 03) 456
3½ m SW of Macclesfield on A536 turn E at Warren for Gawsworth
SJ 8969 (OS 118)

Open late Mar to end Oct daily 1400-1800
♿ 🅿 WC 🍴 (by appt) ♣ 🐕 ✗ ● (no flash)

120

Little Moreton Hall

Congleton, Cheshire

Little Moreton Hall is probably the best-known example of timber-framed architecture in England. It is a delightful building and full of atmosphere, surrounded by a knot garden based on a design of 1688 and then by a water-filled moat. What strikes the visitor first is the charming crookedness of the house – it is hard to find a vertical or horizontal line anywhere. It was built by three successive generations of the Moreton family, who owned the property from about 1250 to the present century. The bridge over the moat leads to a gatehouse and then to the courtyard around which all the buildings are grouped, the main entrance to the great hall being across the yard. The hall is the oldest part of the house, and was 'modernised' in 1599 by William Moreton, who gave it a fireplace in the north wall and a first floor. This was removed before 1807, but the remains of the beam which supported it are still visible. The rooms on either side, completed in 1480, were improved in the 1550s by the addition of large bay windows. The other sides of the courtyard, including the long gallery at the top, with its crazily tilting floor and odd allegorical plasterwork, were completed about 1580, and the house has hardly been touched since. It is almost empty of furniture now, except for two pieces in the hall and the 'great Rounde table in the parlour', but there is painted decoration, some lovely early panelling and splendid decorated fireplaces in the drawing room and upper porch room. A major programme of restoration work has been going on since 1977.

☎ Congleton (02602) 272018

4 m SW of Congleton on A34

SJ 8358 (OS 118)

Open Mar and Oct S and Su; Apr to end Sept daily exc Tu and Good Fri 1400-1800 (or dusk if earlier)

⊖ P WC ♿ 🚻 (by appt) ♠ ☗ (parties must pre-book) ◆ ✿ ✗ NT Open-air play in July

Lyme Park

Disley, Stockport, Cheshire

A large and imposing stone house set in a huge park, nine miles round. The house, home of the Legh family since Tudor times, and given to the National Trust in 1946, is partly of the 16th and partly of the 18th century. About 1550 Sir Piers Legh replaced an earlier hall with the fine courtyard, the main lines of which, with its early Renaissance gateway, can still be seen. From 1650 to 1720 further improvements and reconstructions were carried out, and in 1725 the architect Giacomo Leoni was brought to Lyme to modernise the house. He built the splendid west front facing the lake, made some alterations to the courtyard to give it its present Italianate look, and improved some of the rooms inside. Some further alterations were made in the early 19th century by Lewis Wyatt, who added the tower above the south front to house the servants. The showpiece of the house is the saloon, with its lovely carved wood decorations believed to be by Grinling Gibbons. This room forms part of Leoni's alterations, as does the fine entrance hall with its Mortlake tapestries. Hunting the stag was the traditional occupation at Lyme, so it is fitting that herds of deer still roam in the great park, which itself still retains much of its old forest character, being set in wild, open country. Various activities, such as nature trails, are suggested to the visitor, and the park is frequently the scene of events in the country calendar.

☎ Disley (06632) 2023

8 m SE of Stockport on A6 turn S at Disley

SJ 9682 (OS 109)

Open Apr, May S 1400-1700; June to Sept T-S 1400-1700, Su 1300-1800; Oct S, Su 1300-1600

⊖ (1½ m) ℗ WC 🚻 (by appt) 🚻 (by appt) D (grounds only) ♠ ♣ ⊼ ◆ ⚒ 𝄡 (T-Th) ● 🕆

Bodrhyddan Hall

near Rhyl, Clwyd

This house is the very old home of the Conwy family who have held the hereditary office of Constable of Rhuddlan Castle, a few miles distant, since 1399. However, the feature which has stamped its character on the whole building dates from only 1874 – the spectacular three-storeyed 'Queen Anne revival' frontage (actually, it is at one side) built by W. E. Nesfield. Bodrhyddan comes at the beginning of an architectural fashion that was to fill the suburbs of Britain with red-brick, rather Dutch-looking, gabled houses, and to inspire some grand country houses down to the period of Edwin Lutyens in the late 1890s. The old house behind this frontage preserves an Elizabethan carved stone doorway as well as the former front door, dated 1696. In the current front hall, the owners display a collection of arms and armour. Another, smaller, room contains two mummy-cases brought back from Egypt in the 1830s by an intrepid lady traveller of the Conwy family, Mrs Rowley, whose portrait hangs in the big dining room. In one of the cases is the embalmed body of a young priest. Beyond the front hall is the old hall, which Nesfield clad in light oak panelling. Two portraits dated 1696 by Jean-François de Troy hang there. The white drawing room upstairs has been altered in colour for its present use from its former character of the library, dark with carved woodwork. The family portraits in the 18th-century big dining room include works by Hogarth, Reynolds and Ramsay.

3 m SE of Rhyl on A525 continue on A5151 for 1½ m

SJ 0478 (OS 116)

Open June-Sept T and Th 1400-1730

⊖ P WC ♿ (limited access) ⌷ ♠ ☂ ☐
✿ ⚲ ● (not in house)

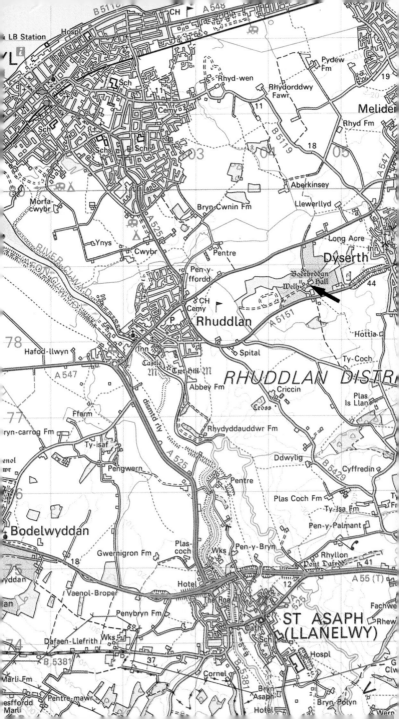

Erddig

near Wrexham, Clwyd

At Erddig all the workings of a large 18th- and 19th-century country estate can be seen in minute detail. The owners, the Yorkes, never threw anything away, and the National Trust, who now own the property, have put it all on show. The large plain, red-brick house was built in the 1680s by Thomas Webb for Joshua Edisbury, and the wings were added in 1724 by the next owner, John Mellor. The Yorke family inherited the property in 1733, and further improvements were made in the 1770s, including the stone facing of the west front. Visitors are shown into the house through the back yard and servants' rooms, and all the outbuildings and below-stairs rooms have their original fittings. They are arranged as though they were still in use, and the servants themselves are immortalised in a series of portraits which hang in the servants' hall and the corridor. These were all commissioned by the family, and some are accompanied by doggerel verses by the kindly but eccentric Philip Yorke II. The main rooms of the house are pleasant but not noteworthy, with the exception of a severe Neo-Classical dining room designed by Thomas Hopper in 1826. The furniture, however, is outstanding. Most of the best pieces were bought in London by John Mellor before he moved into the house in 1716. There are carved and gilded chairs, lacquered and japanned pieces, lovely pier glasses, Soho tapestries and a state bed with Chinese embroidered hangings. Apart from these treasures, the house is crammed with a medley of family possessions.

☎ Wrexham (0978) 355314

1 m SW of Wrexham on A483 turn S to Erddig Park

SJ 3248 (OS 117)

Open Easter to mid Oct daily exc F 1200-1700

⊖ P WC 🖔 (limited access) 🖵 ♣ ⬤☖ ◆
⁂ (braille guide available) ● NT

St David's, Bishop's Palace

Dyfed

Just inland from St David's Head – the westernmost point on the Welsh coastline – stands the village of St David's with its famous cathedral. The settlement grew up initially as a shrine to the patron saint of Wales, but in 1081 William the Conqueror visited it and resolved to incorporate it into the Anglo-Norman ecclesiastical system. As a result, St David's became an English see in 1115. The Bishop's Palace, built a few hundred yards away from the town and cathedral, is now a ruin of imposing proportions, and makes a considerable impact on first viewing. This is mainly thanks to the extraordinary arcading that runs along the parapets in place of the usual battlements of balustrades. The arcading was an idiosyncracy of Bishop Gower, who greatly extended the palace (1327-47), though much of the structure originated before his time. The gatehouse is late 13th century, and leads into a nearly square courtyard round which the buildings are arranged. One small but splendidly vaulted room dates back to *c.* 1200, but the bishop's hall and solar, chapel and gatehouse were all part of an ambitious building programme, undertaken after a royal visit in 1284 showed up the deficiencies of hospitality at St David's. Although the bishop's hall was a sizable building, Gower added a great hall; entertainment must have been lavish, for both halls evidently remained in use. The decline of the palace began when Bishop Barlow (1536-48) removed the lead from the roof (to provide dowries for his five daughters), and by the late 17th century it was uninhabitable.

☎ St David's (0437) 720517

In centre of St David's off A487

SM 7525 (OS 157)

Open daily: mid Mar to mid Oct 0930-1830; mid Oct to mid Mar 0930-1600 (Su 1400-1600)

⊖ P WC ♿ (limited access) ⊟ ♣ ◆
(in summer for groups) WHM

Chavenage

Tetbury, Gloucestershire

Chavenage is mainly Elizabethan; the date 1576 is carved on the porch, although it may have been built on to an existing medieval house. It was built for the Stephens family, and it has always been, and still is, the centre of a working farm, with fine outbuildings. The house is on the common Tudor E-plan, though there have been several later additions, and the garden front is a mixture of every style of architecture from the 16th century to the 20th. The porch has 17th-century Dutch stained glass. The very fine great hall has part of its original screen, a splendid fireplace with the arms of Robert Stephens, who was here until about 1608, and some medieval stained glass in its two tall windows. This probably came from the old Horsley Priory nearby. The drawing room has interesting panelling, some of which is dated 1627, while some, carved with portrait heads or allegorical figures and gilded, is Renaissance in style and seems likely to be mid-16th century. This room also contains some good porcelain. Upstairs there are two 17th-century bedrooms with dark oak furniture and tapestry-hung walls. These are called Cromwell and Ireton after the two generals who stayed here in 1648. Nathaniel Stephens was a staunch Parliamentarian, and Chavenage had been a headquarters during the Civil War. The exit is through the back part of the house, which was enlarged in 1905 by John Mickelthwaite. The nearby chapel has carved figures set into 17th-century stonework.

☎ Tetbury (0666) 52329

2 m NW of Tetbury on B4014 turn W to Chavenage Green for 1 m

ST 8795 (OS 162)

Open Easter Su and M; May to Sept Th, Su and Bank Hol M 1400-1700; parties by appt, other times

🅿 (limited) WC 🍴 (by appt) ♿ ✗ (compulsory)

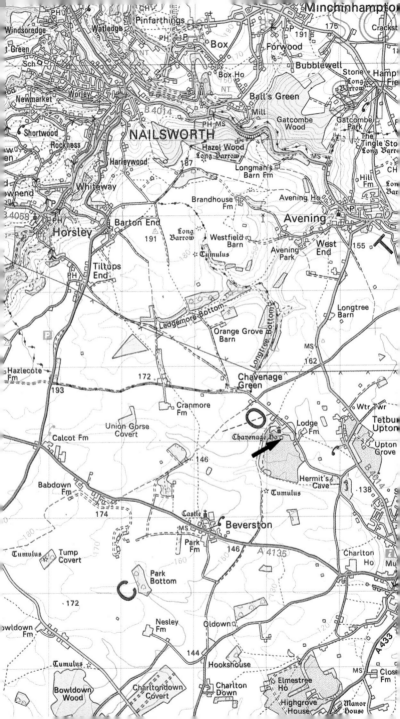

Littledean Hall

Littledean, Gloucestershire

The Hall has been built over centuries; the 'Jacobean' north front is a Victorian replacement of an early Georgian one, which in its turn was a conversion of a genuine Jacobean front. The remains of a Norman house are enclosed within the structure of the north front, and this stands on the walls of a Roman building, which was rebuilt as a church and became a house after the Norman Conquest. Gradually this Norman house grew into a manor house, with a great hall added in the 14th century and further additions made in the 15th. In 1612 Charles Bridgeman bought the house and completely overhauled it to provide a country squire's residence, and in 1664 he sold it to the Pyrke family, who retained it until 1896 and made considerable changes during this period. The interiors are mainly a mixture of the 17th and 18th centuries, but the old parlour, which houses a small museum, gives an idea of an Elizabethan parlour with its huge open fireplace. The overmantel in the library is a fine example of Flemish 16th-century craftsmanship, and the room also has fine 17th-century panelling, as do three of the other rooms. To the right of the fireplace there is a secret closet, and at one time there was a mechanism in the fireplace which allowed the overmantel to swing open, revealing a ladder. The gardens and grounds are open to the public; the Victorian ornamental garden is gradually being replaced by a new water garden with pools stocked with exotic fish.

☎ Dean (0594) 24213

13 m SW of Gloucester on A48 turn W onto A4151 to Littledean

SO 6713 (OS 162)

Open Apr (or Easter if earlier) to Sept daily 1030-1800; parties at other times by appt

⊖ Ⓟ WC ♿ (limited access) 🚏 (by appt)

D (grounds) ♣ ♥ ⌅ ♨ 🏊 (by appt) ⚤

Stanway House

Stanway, Cheltenham, Gloucestershire

This delightful Jacobean manor house was built between 1580 and 1640 on the site of an abbey by the Tracys, a local landowning family. Its interest lies in its development as a typical squire's house, a gradual accumulation through the generations of buildings, furniture and paintings of local significance. The lovely gatehouse ascribed to Timothy Strong of Barrington, was built about 1630 and is adorned with scallop shells, the crest of the Tracys. The south front of the house, built about 1640, linked two existing but separate buildings, the great hall at the west end and the abbot's house at the east end; the curved 'broken pediment' in the middle was added in 1724. An arch to the left of the kitchen range leads past the brewhouse to the entrance porch, and visitors go into a corridor, through the dining room and the former lamp room, where the lamps were cleaned, and then into the audit room. This is where the rents were, and still are, collected from the tenants, and the special 'rent table' with a revolving top made about 1780 is still in use. The other rooms contain a fine collection of family portraits, several good tapestries and some interesting pieces of furniture, notable examples being a shuffleboard table in the hall, made about 1620, and a pair of Chinese Chippendale daybeds in the drawing room, which also contains two pianos by John Broadwood. The buildings outside the main house are equally varied: the tithe barn with its stone roof and oak timbers was built about 1370, and the Pyramid on the hill behind the house in 1750.

☎ Stanton (038 673) 469

10½ m NE of Cheltenham on A46 turn E onto B4077 to Stanway

SP 0632 (OS 150)

Open Jun, Jul, Aug T and Th 1400-1700; parties at other times by appt

⊖ P WC 🖥 ♦ ☕ (by appt for parties) ⚹ ⚔
● (not in house)

Tredegar House

Newport, Gwent

Now one of the finest late 17th-century houses in Britain, Tredegar House was the home of the Morgan family from the early 15th to the 17th centuries. The original medieval stone house was apparently quite substantial, and William Morgan completely rebuilt it between 1664 and 1672 in brick. The new house is very grand, and the stable block in front, built 1717-31, is no less so, while the wrought-iron gates at right-angles to the stable block, made between 1714 and 1718 by William and Simon Edney, are among the finest in the country. The Morgans left Tredegar in 1951 and most of the contents were sold, but it was bought by Newport Borough Council in 1974 and is now being restored and refurnished. Three of the ground-floor rooms, the side hall, dining room and morning room, which were altered in the 19th century, are being refurnished and decorated as Victorian rooms, while the state rooms, which occupy the main front, are 17th century. The best of these are the brown room, which has lavish carved panelling and a plaster ceiling in the 17th-century style, and the gilt room, with walls inset with paintings, gilt decoration gleaming from every surface and an elaborate 17th-century gilt and stucco ceiling. The carved and decorated 17th-century staircase leads to panelled bedchambers, after which a back staircase leads to the servants' quarters. The wine cellar has a 'bath' at one end for washing bottles and cooling wine, which is fed by a natural spring running beneath the house.

☎ Newport (0633) 62275

SW of Newport at junction of A48 and B4239

ST 2885 (OS 171)

Open Good Fri to end Sept W-Su and Bank Hols 1230-1630; parties at other times by appt

 ⊖ P WC ♿ ☐ D ♣ (open all year) ☕
◆ ❄ ✗ (compulsory) ● ♿

Berrington Hall

Leominster, Herefordshire

Berrington is a rectangular Classical mid-Georgian house, built of the local pinkish sandstone, with a large and graceful columned portico. It was built between 1778 and 1781 for Thomas Harley, then Lord Mayor of London, who chose the site with the help of Capability Brown, and the architect was Henry Holland. It is a perfectly proportioned building, its austerity enlivened by the unusual variety of window and door shapes on the front, and the servants' hall, kitchens and other offices are grouped round a courtyard at the back so that the symmetry is unbroken. If the exterior is rather severe for some tastes, the interior certainly is not: the decoration throughout is of an elaborate richness unparalleled in Holland's other work, and many of the colour-schemes and furnishings are original. The entrance hall, with polychrome marble paving echoing the pattern of the ceiling, leads to a magnificent staircase hall, a brilliant exercise in spatial design and still in its original clear, pale colours delicately picked out with gilding. The most splendid of the actual rooms is the drawing room, which has a marble chimneypiece, and one of the most elaborate ceilings Holland every designed. The central part is probably the work of Biagio Rebecca, who was also responsible for some of the decoration in the library and business room. The boudoir, which opens out of the drawing room, is also very fine, with columns of lapis-lazuli blue and a ceiling picked out with pink, blue and gold. The outbuildings include a Victorian laundry and the original dairy.

☎ Leominster (0568) 5721

5 m NE of Leominster on A49 turn W to Moreton

SO 5163 (OS 149)

Open Apr, Oct S, Su, Bank Hol M 1400-1700; May to end Sept W-Su and Bank Hol M 1400-1800

🅿 WC ⊟ ♣ ⬤ ⍨ ◆ ⚘ ⚓ (by appt)
⬤ (no flash or tripods) NT

Cwmmau Farmhouse

Brilley, Whitney-on-Wye, Herefordshire and Worcestershire

Cwmmau farmhouse is a notable case of the mixed blessing resulting from remoteness of setting. Situated in a peaceful rural landscape, it stands high on a ridge close to the Anglo-Welsh border, looking right out across the Wye Valley. Historic events and industrial developments have passed it by – so much so that very little is known about the history of this interesting old house. On the other hand, 'progress' has not spoiled it: despite its workaday character, Cwmmau has not been altered out of recognition, but remains a rare example of an early 17th-century farmhouse, with such original features as its two-storey porch and stone-tiled floors. It is still a family home, and still the centre of a small, self-supporting family farm. Cwmmau – pronounced 'cooma' – is a Welsh word meaning 'many dingles.'

☎ Clifford (04973) 251

3 m S of Kington on A4111, turn SW

SO 2851 (OS 137)

Open Easter to end Aug Bank Hol weekends S-M 1400-1800; by appt at other times

● (no flash or tripods) NT

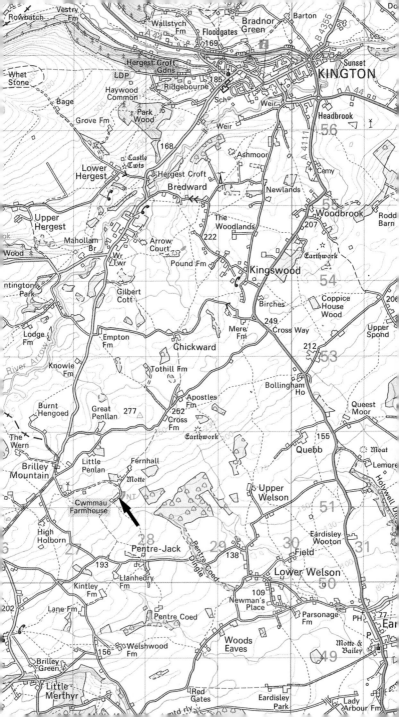

Attingham Park

Shrewsbury, Shropshire

Looking rather like a Classical cube dumped in the featureless countryside that was the despair of the garden designer, Humphry Repton, Attingham has an interior of considerable elegance and distinction. The house was put up in 1783-85 to envelop a house of around 1700. The owner was the Whig magnate Noël Hill and the new place was to go with his title (1784) of Lord Berwick. It is the only surviving country house of the Scottish architect, George Steuart. The layout of the very fine rooms, and the delicately painted wall decorations, show the influence of French taste as was often the case with the houses of the Whig oligarchy in the years before the French Revolution. Opening off the hall on the left are the baron's apartments with the dining room, and on the right, Lady Berwick's with the drawing room. The painted and moulded ceilings recall the Adam style, and the octagon room and the round boudoir both display a variety of touches in the Neo-Classical manner. The second Lord Berwick, returning from a Grand Tour in his twenties, engaged Repton to improve the park, and then, in 1807, brought in John Nash to construct the picture gallery. This is one of the earliest structures in the country to be lit from above by a glass roof supported on a cast-iron framework. The collection of paintings is still hung there in close order all the way up the wall, as was usual in the 19th century. The contents of Attingham were sold in 1827, but the 3rd Baron was able to replenish the house with his own collection.

☎ Upton Magna (074 377) 203

4½ m SE of Shrewsbury on A5 turn N for ½ m at Atcham

SJ 5410 (OS 126)

Open Apr to end Sept S to W; Oct S, Su 1400-1730; Bank Hol M 1130-1730; grounds open all year

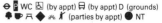

➌ P WC ♿ (by appt) 🚻 (by appt) D (grounds)
♣ 🍴 ⛩ ◆ ⚘ ♪ (parties by appt) ● NT

136

Weston Park

Weston-under-Lizard, near Shifnal, Shropshire

This complex of red-brick and stone buildings – the house, home farm, stables and church – set in an enormous park, more or less forms a village. The large brick house, built in 1671 for Sir Thomas and Lady Wilbraham, is an early example of the Classical style designed by Lady Wilbraham herself, who was a follower of the ideas of the Palladians. In the 18th century the estate passed to the Bridgeman family, later Earls of Bradford, who made alterations and improvements to house and park, and inside the house nothing survives from the earlier period. Many of the rooms have 18th-century marble chimneypieces and plasterwork, but there have been three major remodellings, and the great dining room with its stuccoed ceiling was only completed in 1968, in time for the coming-of-age of Viscount Newport. Two rooms, however, remain entirely 18th century: the charming library with its grained woodwork and the tapestry room, hung with a set of 18th-century Gobelin tapestries. There is a very fine collection of paintings; as well as Dutch and Italian paintings and works by Gainsborough, Reynolds and Stubbs, there are some interesting small portraits, including one by Holbein of Sir George Carew, several recently cleaned Van Dycks and, perhaps most important, Jacopo Bassano's *Way to Golgotha*, which has pride of place in the hall. The park has several attractive buildings, including a Roman bridge and a Temple of Diana built by James Paine about 1760, and behind the temple is a large adventure playground.

☎ Weston-under-Lizard (095 276) 207/385
9 m E of Telford on A5 turn S at Weston-under-Lizard
SJ 8010 (OS 127)

Open mid Apr to end May, also Sept S, Su and Bank Hols; June to mid July daily exc M, F; mid July to end Aug daily; 1100-1700

🅿 WC ♿ 🍴 (by appt) D ♣ ♬ ◆ ⚘ 🕺 🎿 🧍

Cheshire

Arley Hall, Northwich, Cheshire (tel Arley [056 585] 353). 5 m N of Northwich, turn SW. SJ 6780 (OS 109). Early-Victorian 'Jacobean' house and private chapel, with extensive and varied gardens, including walled garden, topiary, herbaceous borders, woodland walk and animals. The house has a collection of 19th-century water-colours of local great houses. Open Apr to Oct T to Su pm, and Bank Hols pm. ▯ WC ⊠ ⊟ (by appt) D (not in house) ♣ ◆ ⊓ ◆ ⊛ ⊁ ⊀ (not Su) ● (not in house)

Capesthorne, Macclesfield, Cheshire (tel Chelford [0625] 861221). 7 m S of Wilmslow on A34. SJ 8472 (OS 118). House built in the 1720s and rebuilt in the mid-19th century in the Jacobean style for the Davenport family, who have lived on the site since Norman times. The interior contains 18th- and 19th-century furniture, paintings, family muniments and Americana. The chapel of 1722 is the earliest known work of John Wood the Elder. There is an extensive park. Open June to Aug T, W, Th, S and Su pm; May and Sept T, Th and Su pm; Apr Su pm only; also Bank Hol M. ⊖ (limited) ▯ WC ⊠ ⊟ D (park only) ♣ ⊡ ◆ ⊁ ⊀ (by appt) ● (not in house) ⊁ ⊀

Churche's Mansion, Nantwich, Cheshire (Nantwich [0270] 625933). In town centre, at junction of A534 and A51. SJ 6552 (OS 118). Elaborately decorated half-timbered house built in the 1570s on the H-plan, for Richard Churche, a local merchant. Restored in the 1930s, and now housing a restaurant, with upper floor open for visitors, showing details of house construction. Open Apr to Oct am and pm. ⊖ ▯ WC ⊟ ⊡ ◆ ⊁

Dorfold Hall, near Nantwich, Cheshire (tel Nantwich [0270] 625245). 1 m W of Nantwich on A534. SJ 6352 (OS 118). Red-brick gabled Jacobean house with diaper-pattern decorations. The great chamber has fine original plaster-work and panelling; other rooms have 18th-century furniture and decorations. Open Apr to Oct, T and Bank Hol M pm; other times by appt. ⊖ ▯ ⊟ (by appt) ♣ ⊁ ⊀ (compulsory)

Handforth Hall, Handforth, near Wilmslow, Cheshire. ½ m E of Handforth on B5358. SJ 8683 (OS 109). 16th-century half-timbered manor house built for the Brereton family, with fine Elizabethan oak staircase and furniture. Open June to Sept by appt. ⊖ ▯

Peover Hall, Over Peover, Knutsford, Cheshire (Lower Peover [056 581] 2135). 3 m S of Knutsford on A50, turn E. SJ 7773 (OS 118). Stone Elizabethan house of 1585, built for Sir Ralph Mainwaring, with a notable kitchen and 17th-century stables with original stalls. There is an Elizabethan garden with topiary, summer-house and lily pond, and an 18th-century landscaped park. The hall was used by General Patton as American 3rd Army HQ in 1944. Open May to Sept M pm (hall, stables and gardens); Th pm (stables and gardens only). ▯ WC ⊠ (limited access) ⊟ (by appt) ♣ ⊡ ⊁ ● (not in house)

Vale Royal Abbey, Whitegate, Northwich, Cheshire (Sandiway [0606] 888684). 2 m NW of Winsford on A54, turn N. SJ 6369 (OS 118). Cistercian abbey founded 1281, but rebuilt as a private house in the 16th century. The interior is presently undergoing extensive restoration. Attractive grounds, with the nun's grave, an historic monument. Open S, Su and Bank Hols, am and pm. ▯ WC ⊠ (limited access) ⊟ (by appt) D ♣ ⊡ ⊁ ⊀

Mid Glamorgan

Llanharan House, Llanharan, Mid Glamorgan (tel Llantrisant [0443] 226253). 7m E of Bridgend on A473. ST 0083 (OS 170). Mainly Georgian country house with notable plasterwork, furniture and and portraits, and a distinctive spiral staircase. There is a fine garden, with hill walks from the house. Open by appt only. ⊖ 🖩 🖵 (by appt) ♠ 🖝 (for parties by appt) ◆ 𝒳 ●

Gloucestershire

Arlington Mill, Bibury, Gloucestershire (tel Bibury [028 574] 368). 7 m NE of Cirencester on A433. SP 1106 (OS 163). 17th-century corn mill situated on River Coln. The interior contains a museum of working machinery, agricultural implements, Staffordshire pottery, Victorian costume etc. Open mid March to mid Nov daily am and pm; Nov to Mar S and Su am and pm only. ⊖ 🖩 WC 🖵 ♠ ⊓ ◆ ⚹ 𝒳 (by appt)

Buckland Rectory, near Broadway, Gloucestershire (tel Broadway [0386] 852479). 2 m SW of Broadway on A46, turn SE. SP 0836 (OS 150). The oldest surviving rectory in the county, built in the late 15th century with 17th-century additions. The house has associations with John Wesley. The hall has a hammerbeam roof. Open May to Sept M am and pm; also Aug F am and pm. 🖩🖵 (by appt) D ★ ● (with permission)

Owlpen Manor, near Dursley, Gloucestershire. 6 m SW of Stroud on B4066, turn E. ST 7998 (OS 162). Medieval stone-built house, with a hall and chamber of the mid-16th century and a Jacobean wing. Open by appt only. 🖩

Painswick House, Painswick, Gloucestershire (tel Painswick [0452] 813646). 6 m S of Gloucester on B4073. SO 8610 (OS 162). Palladian-style country house, built in the 1730s, and enlarged a century later. The drawing room contains 18th-century Chinese wallpaper. Open to groups by appt at any time; also Rococo gardens June to Sept W and 1st Su in month pm. ⊖ 🖩 WC 🖾 🖵 🖝 (by appt) ⚹ 𝒳

Gwent

Llanfihangel Court, Abergavenny, Gwent (tel Abergavenny [0873] 890 217). 6½ m NE of Abergavenny on A465, turn E. SO 3220 (OS 161). Stone-built medieval manor house rebuilt in the 16th and 17th centuries. There is fine mid 16th-century panelling and plasterwork, and an exceptional mid 17th-century yew staircase. There are various outbuildings of similar date. Open July and Aug Su pm; also Easter and Summer Bank Hol Su and M pm. ⊖ 🖩 WC 🖵 (by appt) ♠ 🖝 ⊓ ⚹ 𝒳 ●

Trewyn, Abergavenny, Gwent (tel Crucorney [0873] 890 541). 6 m N of Abergavenny on A465. SO 3322 (OS 161). William-and-Mary mansion built on a medieval site, undergoing restoration. There is notable oak panelling, staircase and minstrels gallery. The grounds include varied ornamental gardens, walks and an aboretum. House open by appt only (closed M); gardens open March to Oct daily (exc M) am and pm. ⊖ 🖩 WC 🖵 (by appt) D (on lead, grounds only) ♠ 🖝 ⊓ ◆ 𝒳 (compulsory) ⚿

Gwynedd

Aberconwy House, Castle St, Conwy, Gwynedd (tel Conwy [0492 63] 2246). In town centre. SH 7877 (OS 115). 14th-century house, the oldest in Conwy. It contains the Conwy museum, covering the history of the town from Roman times. Open Easter to end Sept daily (exc T) am and pm; Oct S and Su only, am and pm. ⊖ 🖵 (by appt) ◆ ⚹ ● NT

Bryn Bras Castle, Llanrug, near Caernarfon, Gwynedd (tel Llanberis [0286] 870 210). 6 m E of Caernarfon off A4086, turn S. SH 5462 (OS 115) 18th-century country house reworked in the early Victorian period in the style of a medieval castle. The gardens are especially extensive and peaceful, with fine views. Open mid July to end Aug daily exc S am and pm; Spring Bank Hol to mid July and Sept pm only. ⊖ (½ mile) 🅿 WC 🚻 (by appt) ♣ 🍴 🎠 ☂

Plas Mawr, High St, Conwy, Gwynedd (tel Conwy [0492 63] 3413). In town centre. SH 7877 (OS 115). Sometimes described as the finest Elizabethan town house in Britain, built by Robert Wynne in the Dutch style between 1576 and 1595. It has been restored to its original condition, with fine furniture and plasterwork ceilings. There is a cockpit in the courtyard. Open daily am and pm (exc Christmas and New Year). ⊖ 🅿 WC 🚻 D ☂

Plas Newydd, Isle of Anglesey, Gwynedd (tel Llanfairpwllgwyngyll [0248] 714795). 1 m S of Llanfairpwllgwyngyll on A4080. SH 5269 (OS 115). 18th-century house built by James Wyatt overlooking the Menai Strait, and decorated in Gothick style. In the dining room is a large fantasy mural painted in the 1930s by Rex Whistler, and there is a room devoted to a Whistler Museum. There is also an Anglesey Museum, with objects related to the 1st Marquess of Anglesey, wounded at Waterloo. The gardens are well known for their sweeping lawns and spring flowers. Open Apr to Sept daily (exc S) pm, Oct F and Su. ⊖ 🅿 WC 🅿 🚻 (by appt) ♣ 🍴 🎠 ◆ ☂ ● NT

Hereford and Worcester

Almonry, Vine St, Evesham, Hereford and Worcester (tel Evesham [0386] 6944). In town centre. SP 0344 (OS 150). Stone and half-timbered house, once part of a

140

14th-century abbey, now containing a museum of local history for the Vale of Evesham. Open Good Fri to end Sept T, Th-Su am and pm (Su pm only); also Bank Hols am and pm. ⊖ 🅿 WC 🅿 (limited access) 🚻 ♣ ◆ ☂

Burton Court, Eardisland, Hereford and Worcester (tel Pembridge [054 47] 231). 5 m W of Leominster on A4112, turn N onto B4457. SO 4257(OS 149). 18th-century house adapted in the Victorian period, but with a 14th-century great hall with a complicated sweet-chestnut timber roof. There is also a collection of working period fairground amusements, and of European and oriental costumes. Open end May to mid Sept W, Th, S, Su and Bank Hols pm. ⊖ 🅿 WC 🅿 🚻 D (not in house) ♣ 🍴 🎠 ☂ 𝄢

Commandery, Sidbury, Worcester, Hereford and Worcester (tel Worcester [0905] 355071). In city centre. SO 8555 (OS 150).An early medieval hospital for the poor, turned into a private house in 1540; the surviving buildings date from the mid-15th century. The great hall has some 15th-century painted glass. The house contains a display of local history, with particular reference to the Civil War; the house formed the local Royalist headquarters in 1651. Open daily am and pm (Su pm only). ⊖ WC 🅿 (limited access) 🚻 (by appt) ♣ 🍴 ◆ ☂ 𝄢

Dinmore Manor, near Hereford, Hereford and Worcester (tel Hereford [0432] 71322). 6 m N of Hereford on A49, turn W. SU 4850 (OS 149). Originally a property of the Knights Hospitallers of St John of Jerusalm, whose 12th- to 14th-century chapel survives; the house, which includes a music room and cloisters, is mainly 20th-century 'medieval'. There is a fine rock garden. Open daily (exc Christmas) am and pm. 🅿 WC 🅿 🚻 ♣ ☂

Hanbury Hall, near Droitwich, Hereford and Worcester (tel Hanbury [052 784] 214). 3½ m E of Droitwich on B4090, turn N. SO 9463 (OS 150). Brick-built house of around 1700, in the style of Christopher Wren and following the model of the nearby Ragley Hall. The staircase and long room are painted by Thornhill; the furnishings are of the 18th century. There is a collection of English porcelain figures, and an orangery. Open May to Sept W-Su, and Bank Hol M, pm; Apr and Oct S, Su, Easter M pm. ⓟ WC 🔄 🚻 ♠ ☛ ◆ ⚹ ⚔ (by appt) ● (no flash or tripods) NT

Hartlebury Castle, near Kidderminster, Hereford and Worcester (tel Hartlebury [0299] 250410 – castle – and 250416 – museum). 5 m S of Kidderminster on A449. SO 8371 (OS 138). The home of the bishops of Worcester for over 1000 years, destroyed in the Civil War and rebuilt in 18th-century Gothick style. The medieval hall is intact. The north wing houses the county museum, including country crafts and industries, costumes, toys, forge, etc. Castle open Easter to 1st Su Sept first Su in the month and W pm; also Bank Hol weekends Su-T pm; parties by appt. Museum open March to Oct daily (exc S) pm; Bank Hols am and pm. ⊖ ⓟ WC 🔄 (limited access) 🚻 (by appt) ♠ 🚻 ◆ ⚹ ⚔ (for parties)

Harvington Hall, near Kidderminster, Hereford and Worcester (tel Chaddesley Corbett [056 283] 267). 4 m SE of Kidderminster off A448, turn N. SO 8774 (OS 139). Moated medieval hall modified in the 1570s and in the early 18th century. The house was owned by the Pakingtons, a Catholic family, and contains a chapel, seven priest holes, and fine Elizabethan murals. Open Easter to Oct daily (exc F) am and pm; Oct, Nov, Feb to Easter pm only. Parties at other times by appt.

ⓟ WC 🔄 (limited access) 🚻 (by appt) ♠ ☛ 🚻 ◆ ⚹ ⚔ (by appt)

Hellen's, Much Marcle, near Ledbury, Hereford and Worcester (tel Much Marcle [053 184] 668). 9 m NE of Ross-on-Wye off A449. SO 6633 (OS 149). Jacobean brick-built manor house with some older elements. There is a collection of 19th-century coaches and old cider presses. Open Good Fri to end Sept W, S, Su pm, Bank Hol M. ⓟ WC 🚻 (by appt) D (not in house) ♠ 🚻 🚻 ⚔ ● (not in house)

Kentchurch Court, Hereford, Hereford and Worcester (tel Golden Valley [0981] 240228). 12 m NW of Monmouth on B4347, turn E. SO 4225 (OS 161). Fortified manor house rebuilt by John Nash. The original gateway survives, and there are distinctive wood-carvings by Grinling Gibbons. Open all year, parties only, by appt. ⓟ WC 🚻 (by appt) ☛ (by appt) ⚔ ● (not in house)

Little Malvern Court, Little Malvern, Hereford and Worcester (tel Malvern [068 45] 4580). 5 m S of Great Malvern on A449. SO 7640 (OS 150). Part of the 12th-century priory of Little Malvern, with Elizabethan and 19th-century elements. The prior's hall is a fine 14th-century room, with exposed timber roof. There is a collection of early needlework. Open mid Apr to end July W, Th pm, other times in same months by appt. ⓟ 🚻 (by appt) ♠ (for parties) ⚹ ⚔

Lower Brockhampton House, Bromyard, Hereford and Worcester. 2 m E of Bromyard on A44, turn N. SO 6855 (OS 149). Small manor house surrounded by a moat, and with a small timber-framed gatehouse, of the 15th century. There is a ruined 12th-century chapel. Open Apr to Oct W-S and Bank Hols am and pm, Su pm; other times by appt. ⓟ 🔄 (limited access) 🚻 (by appt) ● (no flash or tripods) ⚔ NT

Moccas Court, Moccas, Hereford and Worcester (tel Moccas [098 17] 381). 18 m W of Hereford on B4352, turn N. SO 3543 (OS 161). Small red-brick country house built in the late 1770s for the Cornewall family, with details by Robert and James Adam. The interior is elegant, and the house overlooks the River Wye, across a park designed by Capability Brown. Open Apr to Sept Th pm; parties at other times by appt. 🅿 WC 🚻 (by appt) D (not in house) ♦ 𝄞

Old House, High Town, Hereford, Hereford and Worcester (tel Hereford [0432] 68121 ext 207). In town centre. SO 5140 (OS 149). Jacobean house of 1621, once one of a row of timber-framed houses, now restored as a private house with 17th-century furnishings, including much oak furniture. Open M-S am and pm (M am only); closed S pm in winter. ⊝ 🅰 (limited access) 🚻 ♦ ⚬ 𝄞 (by appt)

Tudor House, Friar Street, Worcester, Herefordshire and Worcestershire (tel Worcester [0905] 20904). In city centre. SO 8555 (OS 150). 16th-century timber-framed building, overhanging the street. It houses a museum of local life, with period settings (notably relating to the Victorian age, and to life in the Second World War) and a display of farming implements. Open all year M-W, F-S am and pm; closed Christmas and New Year. ⊝ 🅿 🅰 (limited access) 🚻 ★ ♦ 𝄞 (by appt)

Witley Court, Great Witley, Hereford and Worcester. 5 m SW of Stourport on A451. SO 7664 (OS 150). Ruins of a grand country house, destroyed by fire in 1937, but originally built for Lord Ward in the 1860s on an older house with a portico by John Nash. The terraced gardens include a huge equestrian statue of Perseus by James Forsyth. The chapel, built in the 1730s, is now Great Witley parish church. Open at any reasonable time. 🅿 🚻 D (on lead, grounds only) ♦ ★ EH

Powys

Tretower Court, near Crickhowell, Powys (tel Brecon [0874] 730 279). 3½ m NW of Crickhowell off A40. SO 1821 (OS 161). Remarkable 14th- and 15th-century fortified manor house by the ruins of an earlier cylindrical castle keep. There is a 15th-century great hall, and a small gatehouse. Open daily am and pm; Su pm only). ⊝ 🅿 WC 🅰 🚻 ♦ 🅿 ⚬ 𝄞 (audio) WHM

Shropshire

Acton Round Hall, Bridgnorth, Shropshire (tel Morville [074 631] 203. 7 m NW of Bridgnorth on A458, turn W. SO 6395 (OS 138). Queen Anne period house built for Sir Whitmore Acton of red brick, preserved with its original decorations. The house was abandoned for 200 years to 1918, and now contains a varied collection of furniture belonging to the present owners. Open June to Sept Th pm. ⊝ (1 mile) 🅿 🚻 (by appt) ♦ 𝄞

Adcote, Little Ness, near Shrewsbury, Shropshire (tel Baschurch [0939] 260202). 7 m NW of Shrewsbury off A5. SJ 4119 (OS 126). Victorian country house designed by Norman Shaw in the 1870s, in a style based on Elizabethan houses. There are several tiled fireplaces by William de Morgan, stained glass by William Morris and a large great hall. The house is now used as a school. Open end Apr to mid-July, mid to end Sept, daily pm. 🅿 WC 🚻 (by appt) ♦ ⬤ (by appt) 🅿 ★ (donations requested) 𝄞 ⚲

Benthall Hall, Much Wenlock, Shropshire (tel Telford [0952] 882254). 4 m NE of Much Wenlock on B4375, turn NE. SJ 6502 (OS 127). Stone-built 16th-century house with mullion windows. The interior has 16th-century panelling and 17th-century furniture; there is an intricately carved oak staircase of about 1610. Open Easter to Sept T, W, Su and Bank Hols pm. ⊖ (1 mile) ▯ WC ⚿ (limited access) ⊟ (by appt) ♠ ⚹ ⚔ (by appt) ● (with permission) NT

Boscobel House, near Shifnal, Shropshire (tel Wolverhampton [0902] 850244). 6 m NW of Wolverhampton on A41, turn N. SJ 8308 (OS 127).Small 17th-century timber-framed house in which the future King Charles II took refuge after his defeat at the battle of Worcester in 1651. The rooms that may have been used by him are on view. There is a formal 17th-century garden, and nearby is an oak tree on the site of the Royal Oak in which Charles is said to have hidden. Open daily am and pm (Oct to Mar Su pm only). ▯ WC ⊟ D (on lead, grounds only) ♠ ◆ ⚹ EH

Condover Hall, Condover, Shropshire. 5 m S of Shrewsbury on A49, turn SE. SJ 4904 (OS 126). Late 16th-century manor house, on the E-plan; a fine example of the art of the Elizabethan mason, Walter Hancock, with some 18th-century additions. The house is now owned by the Royal National Institute for the Blind. Open by appt only, Aug pm. ⊖ ▯ ⚿ ♠

Dudmaston, Quatt, Bridgnorth, Shropshire (tel Quatt [0746] 780866). 4 m SE of Bridgnorth on A442. SO 7488 (OS 138). Late 17th-century house built for the Wolryche family, containing fine contemporary furniture. There is a wide-ranging collection of art, particularly flower paintings (including some that belonged to Francis Darby of Coalbrookdale) and 20th-century paintings and sculpture. The house is set in extensive grounds. Open Apr to Sept W and Su pm; parties Th pm by appt. ⊖ (W only) ▯ WC ⚿

(limited access) ⊟ (by appt) ♠ ⚑ ◆ ⚹ ⚔ (for parties by appt) ● (with permission)

Shipton Hall, Much Wenlock, Shropshire (tel Brickton [074 636] 225). 8 m SW of Much Wenlock on B4378. SO 5693 (OS 137). Attractive Elizabethan manor house in a large park. The interior was modified with a new hall and staircase in the 1760s, when a stable block was also built. There is a stone walled garden, medieval dovecote, and the parish church is also in the grounds. Open July and Aug Th and Su pm; May, June and Sept Th and Bank Hol Su and M only; parties at other times by appt. ▯ WC ⊟ (by appt) ♠ ⚑ (for parties by appt) ⚔

Upton Cressett Hall, Bridgnorth, Shropshire (tel Morville [074 631] 307). 4 m W of Bridgnorth on A458, turn SW. SO 6592 (OS 138). Medieval timber-framed manor house modified in the Elizabethan era, displaying original timberwork in the upper rooms. There is also an Elizabethan gatehouse and a medieval garden with peacocks. A Norman church stands in the grounds. Open May to Sept Th pm. ▯ WC ⊟ (by appt) ♠ ⚔

Wilderhope Manor, Wenlock Edge, Shropshire (tel Longville [069 43] 363). 9 m SW of Much Wenlock on B4371, turn S for 1¼ m then NE. SO 5492 (OS 138). Limestone-built Tudor manor house, similar in appearance to Shipton Hall. There are 17th-century plaster ceilings, and the house is set in deep woods. It is now used as a Youth Hostel. Open all year S pm, also W pm Apr to Sept. ▯ WC NT

Central England

W YORKS

GREATER MANCHESTER
Manchester

SOUTH YORKSHIRE

Sheffield

CHESHIRE

LINCOLN-SHIRE

🏠 Chatsworth

🏠 Thoresby Hall

Haddon Hall 🏠

🏠 Hardwick Hall

Carlton Hall 🏠

DERBYSHIRE

NOTTINGHAMSHIRE

🏠 Newstead Abbey

Stoke-on-Trent
🏠 Ford Green Hall

🏠 Whitmore Hall

STAFFORDSHIRE

Ednaston Manor

🏠 Kedleston Hall

Derby

Wollaton Hall 🏠

Holme Pierrepont Hall 🏠
Nottingham

🏠 Sudbury Hall

Melbourne Hall 🏠

🏠 Thrumpton Hall

Stafford
🏠 Shugborough

Hoar Cross 🏠 Hall

🏠 Prestwold Hall

LEICESTERSHIRE

🏠 Hanch Hall

🏠 Donnington-le-Heath Manor House

Chillington Hall 🏠

Belgrave Hall 🏠
Wygston's House 🏠

🏠 Moseley Old Hall

Leicester

Bede House, Lyddington 🏠

Southwick Hall 🏠

🏠 Wightwick Manor

SHROP-SHIRE

Oak House 🏠

🏠 Aston Hall

Kirby Hall 🏠
Deene Park 🏠

Birmingham 🏠 Blakesley Hall
WEST MIDLANDS

🏠 Arbury Hall

Stanford Hall 🏠

Triangular Lodge 🏠

🏠 Boughton House

🏠 Hagley Hall

Coventry

🏠 Kelmarsh Hall
🏠 Lamport Hall

NORTHAMPTONSHIRE

CAMBS

HEREFORD

🏠 Packwood House

🏠 Stoneleigh Abbey

Baddesley Clinton

🏠 Lord Leycester Hospital

🏠 Holdenby House
🏠 Althorp

AND

Coughton Court 🏠

WARWICKSHIRE

1 🏠 2
🏠 🏠 3 🏠 Charlecote Park
4 🏠

Delapré Abbey 🏠
Castle Ashby 🏠

Hinwick House 🏠

Worcester

Ragley Hall 🏠

5

Northampton

WORCESTER

Stratford-upon-Avon
Honington Hall 🏠

🏠 Farnborough Hall

🏠 Canons Ashby House

Chicheley Hall 🏠

BEDFORD-SHIRE

Upton House

🏠 Sulgrave Manor

Aynhoe Park

Banbury

Milton Keynes

🏠 Chastleton House

Winslow Hall 🏠

Ascott 🏠

Ditchley Park 🏠

🏠 Rousham Park

Claydon House 🏠

BUCKS

Mentmore Towers 🏠

Blenheim Palace 🏠

Wotton House 🏠

🏠 Waddesdon Manor

Dorton House 🏠

Nether Winchendon House 🏠

OXFORDSHIRE

Court House, Long Crendon 🏠

HERTS

GLOUCESTERSHIRE

Stanton Harcourt 🏠

Oxford

🏠 Manor House

Chenies Manor House 🏠

Kelmscott Manor 🏠
Buscot House 🏠

Kingston House 🏠

🏠 Nuneham Park

Hughenden Manor 🏠

West Wycombe Park 🏠

Milton's Cottage 🏠

Kingstone Lisle Park 🏠

🏠 Milton Manor

Stonor 🏠

Cliveden 🏠

Swindon

Ardington House 🏠

Grey's Court 🏠

AVON

🏠 Ashdown House

WILTSHIRE

Mapledurham House 🏠

Reading

Dorney Court 🏠

BERKSHIRE

1 Mary Arden's House
2 New Place and Nash's House
3 Halls Croft
4 Anne Hathaway's Cottage
5 Shakespeare's Birthplace

Chenies Manor House

Chenies, near Rickmansworth, Buckinghamshire

Chenies, owned by the Russells until the 1950s, was originally a brick manor house built about 1460 by the Cheyne family, and the central part of today's house, with the tower, is of that date. In 1523 John Russell, 1st Earl of Bedford, married the owner of the manor, and the additions he made, the main one being the long brick range set at right-angles to the old gabled one, gave the house its present appearance. A feature of the house is its ornamental chimneys, very similar to those at Hampton Court, and the two buildings shared some of the same workmen. In the 18th century the north wing became a ruin, and the inner courtyard now has walls on two sides where the domestic buildings once were. The 1st Earl made his extensions primarily to entertain Henry VIII and his Court, and the King made two visits. On the second of these, in 1542, Catherine Howard, the current wife, committed adultery with Thomas Culpeper. It is said that the slow steps of a lame man are sometimes heard leading to Catherine's room – the King was at the time in pain from an ulcerated leg. After 1627 Woburn Abbey became the Russells' main seat, though they preserved Chenies as a relic of the past. The interiors are a mixture of styles: Queen Elizabeth's room contains 16th- and 17th-century furniture, while the dining room was modernised in the early 19th century and contains furniture of that period. Among the Tudor survivals are the privy in a closet off the library, and the small dark closets opening off each of the upstairs rooms.

☎ Little Chalfont (02404) 2888

3 m E of Amersham on A404 turn N

TQ 0198 (OS 176)

Open Apr to Oct: W and Th 1400-1700; parties at other times by appt

⊖ P WC ⊟ (by appt) ♠ ☛ ◆ ❀ ⚶ ✗ ● (not in house) ⟡

Chicheley Hall

Newport Pagnell, Buckinghamshire

Chicheley was built between 1719 and 1723 by Francis Smith of Warwick, a well-known Midlands architect, and its main front, with a projecting and elevated central bay and giant pilasters, is a fine example of the English Baroque style. It is thought that the owner, Sir John Chester, and his friend Burrell Massingberd also had a hand in the design, and contemporary letters suggest that they did not always see eye to eye. Each side of the house is different, and both the highly Baroque doors (which were criticised by Massingberd) and the windows, many of which still have their original glass and thick glazing bars, are important features of the building. The front door, modelled on Bernini's Chapel of the Holy Crucifix in the Vatican, opens directly into an elegant entrance hall designed by Henry Flitcroft in the Palladian style. The early-18th-century staircase, made of oak but with handrail and treads inlaid with walnut, is particularly fine, and was the combined effort of several craftsmen. The main rooms give a comfortable and intimate feeling, and many contain fine panelling and ceilings. A curious feature of the house is Sir John Chester's library, in which the books are entirely hidden behind ingeniously hinged panelling. The present occupants of the house are the descendants of the First World War naval commander Admiral Beatty, and there are many relics of his career in the study, as well as a naval museum on the second floor.

☎ North Crawley [023065] 252

7½ m NE of Milton Keynes on A422

SP 9045 (OS 152)

Open Good Fri to late Sept Su, Bank Hols 1430-1800; parties at other times by appt

▉ WC 🚾 (limited access) 🚻 (by appt) ♠ ⬤ 🎋 ⚒
ⵜ (compulsory for parties)

Dorney Court

Dorney, near Windsor, Buckinghamshire

This delightful, rambling house, of mellow red brick, dates from 1500, but Dorney Court has always been the manor house of the village, and a house on the site was first recorded after the Norman Conquest. The building, with its irregular roofs and gables, set in a traditional English garden with box hedges, seems at first sight to be entirely medieval, but in fact some of the exterior is a Victorian reconstruction. It has been the home of the Palmer family since the late 16th century, and there is a legend that the first English pineapple was grown here by Rose, the gardener, and presented to Charles II in 1661. The interior layout of the house – the relationship of hall to kitchens, buttery, cellar, courtyard and so on – is similar to many medieval houses, and is little changed from 1500. The oldest part of the house is the panelled parlour with its low-beamed ceiling, which sits beneath the great chamber on the floor above (now a bedroom). This room, with its barrel-vaulted ceiling, contains fine furniture, as do the other rooms on display. The great hall, which contains an impressive collection of family portraits and some superb linenfold panelling brought from Faversham Abbey, was at one time used to hold the Manor Court. The only room of a later date from the rest is the dining room, an elegant panelled interior in the William and Mary style.

☎ Burnham (06286) 4638

1 m N of Windsor on B022 turn W onto B3026, after 3 m turn SW at Dorney

SU 9279 (OS 175)

Open Easter to 2nd Su Oct Su and Bank Hol M; June to Sept Su, M and T 1400-1730

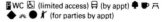

🅿 WC ♿ (limited access) 🚻 (by appt) ♠ 🍽 ⛩ ◆ ❋ ● ⚔ (for parties by appt)

Hughenden Manor

High Wycombe, Buckinghamshire

The house, home of Benjamin Disraeli, Earl of Beaconsfield, for the last thirty-three years of his life, is more interesting from a historical point of view than from an architectural one, and makes an interesting contrast with the nearby Waddesdon Manor, owned by Baron Ferdinand de Rothschild. The owners, both Jewish and both married but childless, were friends but their houses could not be more different, Hughenden being as unassuming as Waddesdon is breathtaking. When Disraeli bought the house in 1847, it was of plain white stucco, dating mainly from the 18th century. In 1862 he and his wife brought in the architect Edward Buckton Lamb to remodel the house in 19th-century Gothic style. Disraeli believed it was 'being restored to what it was before the Civil War', but in fact it simply looks thoroughly Victorian: Lamb faced the outside with red brick and a great many pinnacles and arches, while inside he put in some simple, vaguely Gothic-looking ceilings and chimneypieces. The Disraelis loved Hughenden and spent all the time they could here. Some of the rooms still contain their furniture and pictures, and are arranged as though they still lived in the house, but it is mainly a museum of Disraeli's political career, with letters, mementoes and pictures of all the people he had known during his long life, including Queen Victoria. The Queen was very fond of Disraeli; on one occasion, in 1877, she lunched with him at Hughenden, and when he died she sent a wreath of primroses, 'his favourite flower from Osborne'.

☎ High Wycombe (0494) 32580

1½ m N of High Wycombe W of A4128

SU 8695 (OS 165)

Open Apr to Oct W-S 1400-1800, Su and Bank Hol M 1200-1800; Mar S and Su 1400-1800

⊖ (1½ m) 🅿 WC 🚻 (limited access) 🚻 (by appt) D (grounds only) ♣ 🍴 ⌐ ◆ ✹ NT

Mentmore Towers

Mentmore, near Leighton Buzzard, Buckinghamshire

The splendid Victorian mansion was built for Baron Meyer Amschel de Rothschild in 1855 by Joseph Paxton, the designer of the Crystal Palace. The style is Jacobean, and the house is closely based on Wollaton Hall in Nottinghamshire, using the same stone, and with the same corner towers and decorative features. Mentmore has had an eventful history. In 1878 it became the home of the 5th Earl of Rosebery, and over the next hundred years saw many distinguished guests including Napoleon III, Czar Nicholas II, Winston Churchill and Her Majesty Queen Elizabeth II. In 1977 its entire contents were sold in a much publicised auction, raising £6½ million, and a year later it was purchased by the Maharishi Mahesh Yogi and is now the headquarters of the Maharishi University of Natural Law. The palatial interiors, a reminder of the wealth and power of the Rothschild family, are a mixture of the 'Versailles' and the 'Italian palazzo' styles, with much use of carving and gilding. The rooms are all arranged round a huge central hall surrounded by plate-glass doors and windows, which express Paxton's love of glass and light. The fireplace, of black-and-white marble, was imported from Rubens' house at Antwerp, while the dining room contains ornament from the Hôtel de Villars in Paris. The rooms have recently been redecorated, and the tours, which are conducted by a member of the university, end with a visit to the old servants' wing, now laboratories where research is being conducted into higher states of consciousness.

☎ Cheddington (0296) 661881

4 m S of Leighton Buzzard on B488 turn W to Mentmore

SP 9019 (OS 165)

Open throughout year Su and Bank Hols 1345-1600

⊖ (1 m) 🅿 WC 🚻 (by appt) 🍴 ♦ 🍽 (parties only) ♨ 🚹 ●

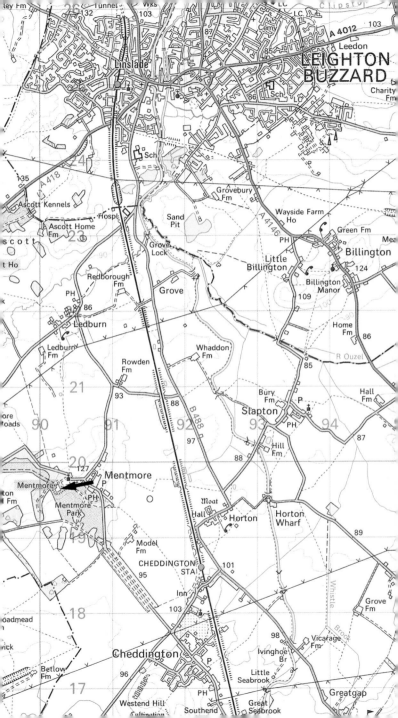

Waddesdon Manor

Aylesbury, Buckinghamshire

Waddesdon is a French château in the heart of the English countryside, but it was built by Baron Ferdinand de Rothschild, who was in love with French 18th-century art. The house was designed in the Renaissance style by Gabriel-Hippolyte Destailleur, and built between 1874 and 1889. It was no small operation, as before building could begin the ground had to be levelled and woods planted on the bare hillside. To speed things up, half-grown trees were dragged to the hilltop site by teams of horses, and a railway was constructed to carry stone to the bottom of the hill. The exterior is surprising but the interior is incredible, illustrating the lifelong passion of this rich and cultured collector. Baron Ferdinand not only had a deep love and understanding of French 18th-century art, he was also a collector of English 18th-century portraits and paintings of the Dutch and Flemish schools, so there are many fine paintings, while rooms upstairs contain more specialised collections of china, musical instruments and costume. Miss Alice de Rothschild, the Baron's sister, was also a keen collector, and added Sèvres and Meissen as well as arms and armour. The best features of the house are the ground-floor rooms, where there are interiors matched only in the great French houses, with incomparable collections of 18th-century furniture, carpets, porcelain and sculpture. Many of these richly decorated rooms contain carved wooden panels (*boiseries*) brought from houses in France.

☎ Aylesbury (0296) 651211/651282

7 m W of Aylesbury on A41

SP 7316 (OS 165)

Open late Mar to Oct W-Su 1400-1800 (exc W after Bank Hol); Good Fri and Bank Hol M 1100-1800

⊖ 🅿 WC ♿ (limited access) 🚌 (by appt)
D (grounds) ♣ ♣ ⚬ 🎋 (by appt) ● NT

West Wycombe Park

West Wycombe, Buckinghamshire

The Dashwood family bought the estate in 1698, and the original brick house was built by Sir Francis soon after, but the 18th-century house we see today owes its character entirely to his son, the second Sir Francis Dashwood, founder member of both the Society of Dilettanti and the Hell Fire Club. From 1769, when he returned from travelling in Italy, he began to alter the house, but it is impossible to fit it into any architectural pigeon-hole as it accurately reflects his ever-changing tastes and sudden enthusiasms. He employed several different architects and artists as well as having a good many ideas of his own, so that the house is a patchwork – albeit a charming one – of 18th-century styles. The Palladian north front, the first to be completed, the south façade with its two-storey colonnade, and the east front with its Doric portico are all the work of a little-known architect, John Donowell. The west front, in contrast, is by Nicholas Revett and has a Greek portico. The interiors are equally mixed in style: the hall and dining room are Neo-Classical, and painted to imitate coloured marbles; the staircase landing, the undersides of the stairs and the ceilings of library and study have delightful Rococo plasterwork, while several of the other rooms have Baroque painted ceilings by an Italian artist, Guiseppe Borgnis. The park, to which Sir Francis devoted as much attention as the house, is attractively laid out and contains several 18th-century 'garden temples'.

☎ High Wycombe (0494) 24411

3 m W of High Wycombe on A40

SU 8294 (OS 175)

Open June M-F, July and Aug daily exc S 1400-1800; grounds also open Easter, Spring Bank Hol

⊖ 🅿WC 🚻 (by appt) ♥ ⚹ 🏃 ● (not in house) 🛉 NT

Winslow Hall

Winslow, Buckinghamshire

This red-brick house, modest in scale for a country house, was almost certainly the work of the great architect Sir Christopher Wren, who built a great many churches but few domestic buildings. It was built between 1699 and 1702 for William Lowndes, Secretary to the Treasury, whose family had lived in Winslow since the early 16th century, and several very well-known craftsmen were employed on the brickwork and carpentry. The building of the house is well documented, and the owners show visitors the original accounts which mention all the craftsmen by name, itemise the various expenses and give a total for the whole work – £6,585. 10s. 2¼d. The exterior is quite simple, with two tall main fronts which are made to look even taller by the four huge chimneys and the pointed central pediment. Originally the flanking buildings for kitchens, servants' accommodation and so on were detached, but in 1901 they were rather clumsily joined to the main house. The interiors are well proportioned, and most of the rooms have their original oak panelling and fireplaces, but with one exception the decoration is pleasant rather than spectacular. The exception is the Painted Room, where four wide spaces on the walls are filled with paintings in heavily ornate frames with scrolls, cartouches and drapery. The artist is unknown, but the paintings were apparently based on a well-known series of tapestries. The gardens were laid out in 1695 but nothing remains of their formal layout, though the present gardens are charming and well kept.

☎ Winslow (029 671) 3433

On NW outskirts of Winslow on A413

SP 7628 (OS 165)

Open July to mid Sept daily exc M 1430-1730; early July, late Sept S, Su; Bank Hol M 1400-1730

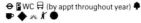 (by appt throughout year)

Chatsworth

Bakewell, Derbyshire

At Chatsworth a wild valley in the Peak district has been tamed to provide an exceptional setting for a magnificent house. The original house was built by Bess of Hardwick in the 1550s; in 1694 William Cavendish, 1st Duke of Devonshire, started rebuilding. By 1707 it had a new face of classic grandeur, the yellow stone embellished by restrained carving. During that period gardens were also laid out. The house was extended but otherwise little altered for the next hundred years, though much was done to the grounds. Capability Brown cleared away most of the formal gardens, the valley was landscaped into beautiful parkland, and plantations reduced the severity of the moors. The 6th Duke inherited in 1811, and spent forty years altering and enlarging the house and its gardens. Jeffry Wyatville built him a new set of rooms for entertaining, and also created a new library, a sculpture gallery and a theatre. When he left, Chatsworth boasted 175 rooms. The superb gardens were made by Joseph Paxton, head gardener at Chatsworth from 1826. He created the arboretum, the pinetum and the rock gardens; rare specimens were brought from all over the world to fill the glasshouses, including the great conservatory that foreshadowed his Crystal Palace. The present Duke and Duchess have made many improvements to the gardens, which are as popular with visitors as the house, itself a treasure and containing many precious objects and a fine collection of furniture, pictures, drawings and books.

☎ Baslow (024 688) 2204

10 m W of Chesterfield on A 619, turn S at junction with A621

SK 2670 (OS 119)

Open 1st Apr to 1st Nov daily 1130-1630

⊖ (limited) 🅿 WC 🖾 ⊟ D (on lead, grounds only) ♣ 🍴 ⊓ ◆ 🕸 (nearby) ✂ (also in foreign languages) ⟋ ● ⳨ ⳨

155

Haddon Hall

Bakewell, Derbyshire

Perched on a limestone slope above the Derbyshire River Wye, the grey walls, battlements and towers of Haddon Hall look like an illustration from a book of medieval history. The Hall was abandoned by its owners, the Manners family, in 1700, but from 1912 was lovingly restored to its former condition by the 9th Duke of Rutland, whose life's work it was. There are two courtyards, divided by the 14th-century range containing the banqueting hall, kitchen and dining room, and the gatehouse leads into the lower court, with a small museum to the left. The chapel contains a remarkable series of medieval wallpaintings, and the roof and woodwork all date from 1624. The roof of the banqueting hall is part of the 1920 restorations, but the screen is of 1450, and the kitchens beyond still contain many of their original fittings. The dining room, completed about 1545, has interesting heraldic panelling, carved medallions in the alcove and heraldic ceiling paintings. Above the dining room, and almost identical in shape, is the great chamber, with its lovely Elizabethan plaster frieze and fine timber roof. The long gallery, a splendidly light and airy room lit by large bay windows, has panelling later in style than the other rooms, with interesting heraldic details. A feature of the bay windows is that some of the leaded lights bulge outwards. This was designed to produce a pretty effect, and was preserved when the gallery was restored. The effect can be seen best from the delightful gardens, which were laid out in terraces in the 17th century, and are now planted with a variety of flowers.

☎ Bakewell (062981) 2855

2½ m SE of Bakewell on A6

SK 2366 (OS 119)

Open Apr to Sept daily exc M 1100-1800; also Bank Hol M; reduced rates for parties

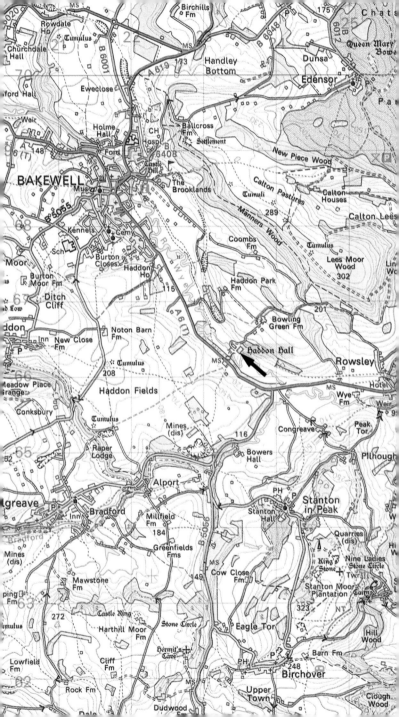

Hardwick Hall

Doe Lea, Chesterfield, Derbyshire

Hardwick Hall, glittering with the glass of its great mullioned windows, and bearing the initials and coronets of its builder on its six towers, is one of the most beautiful Elizabethan houses. It was built by 'Bess of Hardwick', who was born in 1518 in the manor house nearby. She married four times, becoming richer with each new marriage, and Hardwick Hall was built after the death of her last husband, the Earl of Shrewsbury, in 1590. She had already rebuilt her father's old house, which now stands in ruins a stone's throw away. Her architect was Robert Smythson – though she probably made most of the decisions herself – and the building is innovatory in many ways, with the great hall at right-angles to the façade instead of parallel to it, and the state rooms at the top of the house. The private family apartments were on the first floor; the ground floor, apart from the great hall, was mainly used by servants, and the windows become progressively taller to suit this interior arrangement. The hall, which contains pieces of 16th-century appliqué work, leads to the tapestry-hung staircase which ascends past the first-floor drawing room to the showpiece of the house, the high great chamber. It was designed for the tapestries which still line the walls, and above them is a painted plaster frieze by Abraham Smith. Behind this room is the long gallery, in which family portraits are hung over more tapestries. The other second-floor rooms also have rich decoration and fine furniture. The back staircase is hung with more examples of excellent embroidery.

☎ Chesterfield (0246) 850430

6 m NW of Mansfield on A617 turn S at Glapwell

SK 4663 (OS 120)

Open Apr to end Oct W, Th, S, Su and Bank Hol M 1300-1730 (or dusk if earlier)

♿ P WC ⊟ D (on lead, grounds only) ♣ 🐾
𝔸 ◆ ⚘ ● ⚰ NT

158

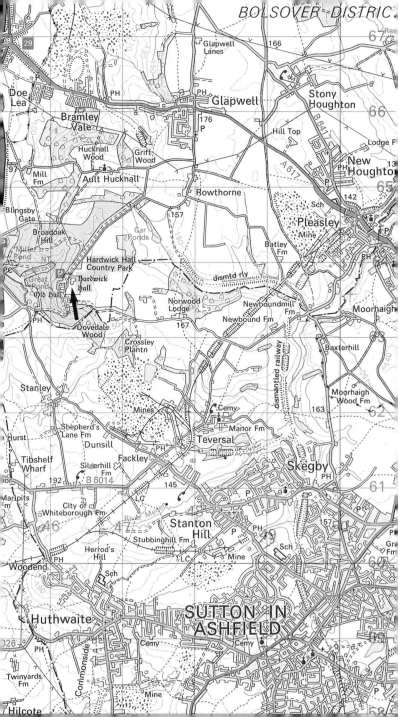

Kedleston Hall

Derby, Derbyshire

Kedleston Hall, home of the Curzons for nearly 900 years, is one of the masterpieces of Robert Adam. It is a vast building, with the great pedimented central block flanked by pavilions big enough to be houses in themselves; but it seems curiously unostentatious. It was built for Sir Nathanial Curzon, 1st Lord Scarsdale, and work began in 1759. The original architect was Matthew Brettingham, who produced the design; James Paine took over in 1760 but was almost immediately replaced by Adam, who designed a completely new south front and modified the entrance front. This south front, with its centre in the form of a Roman triumphal arch and its curving dome and staircase, shows all of Adam's genius, but was never realised, for the money finally ran out. The great portico leads into the marble hall, a stately room which owes more to Brettingham than to Adam. The other rooms were all decorated to Adam's designs, and three rooms on the left of the hall, which were devoted to music, painting and literature, have fine Adam ceilings. The Italian and Dutch paintings here, most of which were bought by Sir Nathaniel, are still arranged as he and Adam placed them. At the centre of the house is the saloon, a circular room based on a Roman temple and designed to display sculpture. Of the four state rooms on the other side of the building, the dining room is the most recognisably 'Adam' in style, and the great apartment is notable for its state bed with posts carved to resemble palm trees.

☎ Derby (0332) 842191

4 m NW of Derby on A52, turn N at Kirk Langley

SK 3140 (OS 128)

House open 1300-1730 Apr to Sept S, Su, M; gardens open 1200; park open 1100

🅿 WC 🚻 ♠ 🍴 ◆ ♨ ● (no flash or video) NT

Melbourne Hall

Melbourne, near Derby, Derbyshire

The parsonage of Melbourne, a small country town, was leased to Sir John Coke in about 1620, and the 17th- and 18th-century Hall, which stands very near the church, may conceal a medieval parsonage. The house is very mixed in style both outside and in, as Sir John, instead of rebuilding, as was usual at the time, merely made some additions to the old house. His son Thomas, who owned the house in the early 18th century, was more interested in the garden and only later considered building a new house. In fact he only remodelled the courtyard and built a new east wing facing the garden. The designer of this garden wing, built in 1725, was the master-builder Francis Smith, but its Palladian front was added by his son William in 1744. The staircase inside, hung with paintings, is a fine example of carved joinery. The original great hall is now a dark-panelled dining room, and most of the other rooms have simple 18th-century decoration. Some of the furniture is very good, and there are some interesting paintings, including several of Stuart courtiers who were friends of Thomas Coke. A fine double portrait by Lely shows Thomas' parents-in-law, Lord and Lady Chesterfield, and there are charming, rather primitive Jacobean portraits of the Cokes' maternal family, the Leventhorpes. The formal gardens have been very little changed since Thomas Coke laid them out with Henry Wise in 1704. The focal point, the wrought-iron arbour or 'birdcage' painted in bright colours and gilded, is an early work by Robert Bakewell of Derby.

☎ Melbourne (03316) 2502

6 m S of Derby on A5132 turn S onto A514 for 1½ m then onto B587

SK 3924 (OS 128)

Open early June to early Oct W 1400-1800; gardens Apr to Sept W, S, Su, Bank Hol M 1400-1800

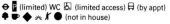
♿ 🅿 (limited) WC ♿ (limited access) 🚌 (by appt) ♣ 🍴 ◆ ⚲ ⚔ ● (not in house)

Belgrave Hall

Church Road, Leicester, Leicestershire

This modest red-brick house, standing on the edge of the suburbs of Leicester, was built in 1709-10 for Edmund Cradock, 'gent', son of a local innkeeper whose family was prominent in local affairs. In spite of its very splendid iron gateway, the exterior of the house is very plain, though not actually ugly. It has passed through various hands since the Cradocks' time, and in 1936 was acquired by the Leicester Museums Service, who have furnished each of the rooms in the style of a particular period from the late 17th century to the 1870s. Although none of the furniture is that which belonged to the actual occupants of the house, the rooms give a very good idea of the lifestyles of moderately well-to-do famies in the 18th and 19th centuries, in contrast to the great country houses, where everything is for show rather than for use. The most outstanding furniture is a set of early 18th-century mahogany chairs with embroidered covers, and there is also a fine red lacquer bureau. These are in the drawing and dining rooms respectively. Other rooms on show are bedrooms, a workroom, a music room, a Victorian nursery and a large kitchen of the early 19th century. There is an attractive garden at the back of the house, and the stables contain an interesting collection of farm implements and harnesses as well as an 18th-century coach.

☎ Leicester (0533) 554100

2 m N of Leicester city centre on A6

SK 5806 (OS 140)

Open throughout year M-S exc F 1000-1730, Su 1400-1730. Closed 25,26 Dec

⊖ ♿ (limited access) 🚻 (by appt) ♣ ★ ◆
🍴 (by appt)

Stanford Hall

Lutterworth, Leicestershire

This large and dignified William and Mary house was designed by the well-known Midlands architect William Smith of Warwick for Sir Roger Cave, whose descendants still live here. It was begun in the 1690s as a replacement for an older house, but both Sir Roger and William Smith died during the building, and it was completed by Smith's sons, William the younger and Francis, for Sir Thomas Cave. The stable block and courtyard were added in the early 18th century by Francis Smith, who was also responsible for many of the interiors. Most of the ground-floor rooms and two of the upstairs are open to visitors, and there is much to see, including a unique collection of Stuart portraits and religious relics. These, which are displayed in the ballroom, were part of the collecton of Cardinal Henry Stuart, and were bought by Baroness Braye after 1807. Most rooms have their original panelling and fireplaces, and there is very good furniture, including Hepplewhite and Queen Anne pieces and a particularly fine 17th-century marquetry clock. The green drawing room has 17th-century portraits, and the marble passage and the stairs are both hung with family portraits. The bedrooms contain 17th-century Flemish tapestries as well as tapestry curtains and bed-hangings, and the old dining room has some furniture from the old house, including an old refectory table and Charles II chairs. The aviation museum in the stables was set up in memory of Percy Pilcher, a pioneer of aviation, who was killed when his machine crashed in the park.

☎ Rugby (0788) 860250

8 m NE of Rugby on B5414 turn E for 1½ m at Swinford

SP 5879 (OS 140)

Open Easter to end Sept Th, S, Su, Bank Hol M and following T pm

 WC 🚻 (by appt) D 🚶 🍽 🎋 ◆ ♒ ⚹
● (not in house) ⚹

Althorp

Northampton, Northamptonshire

Home of the Spencer family, the house was originally built by Sir John Spencer soon after 1508. His red-brick house surrounded by a moat and formal gardens was remodelled in 1650 by Dorothy, wife of the 1st Earl of Sunderland, and the inner courtyard became the saloon and great staircase. The house owes most of its character to the architect Henry Holland, who in 1786 was called in to remodel it again. He refaced it in the fashionable grey-white brick tiles baked at Ipswich, filled in the moat and improved the gardens with the help of Samuel Lapidge, Capability Brown's chief assistant. The grand stable block was designed by Roger Morris and built in 1732. Morris was also responsible, with Colen Campbell, for the splendid lofty Palladian entrance hall with its lovely plaster ceiling. This room is known as the Wootton hall as it was designed for equestrian paintings by John Wootton, which exactly fit the walls. The rooms at Althorp house one of the finest private collections of art in England, which was begun by Robert Spencer, 2nd Earl, and enlarged and improved by successive generations. The Marlborough room contains particularly fine portraits by Gainsborough and Reynolds; there are portraits by Rubens and many Old Masters; the picture gallery has works by Van Dyck, Lely and others, while the staircase is hung with portraits of the Spencer family since the time of Elizabeth I. The porcelain and furniture are equally fine, and there is some marvellous 18th-century furniture.

7 m NW of Northampton on A428 turn SW

SP 6865 (OS 152)

Open daily throughout year; Oct to end Jun 1420-1730; Jul to end Sept 1100-1800

⊖ 🅿 WC 🛏 ♣ 🍴 ◆ ⚘ 🎋 ● (not in house)

Boughton House

Kettering, Northamptonshire

Ralph, 3rd Lord Montagu of Boughton, was sent on several ambassadorial missions to the court of Louis XIV, and so great a taste did he acquire there for French style that he transplanted the idea for this palatial country seat to England. Not only would the exterior look as if it would be at home in the main square of a 17th-century French provincial town, but the friezes and ceilings inside were painted by a French artist, Louis Cheron. Begun in the late 1680s to absorb a straggling Tudor manor built on land bought by an ancestor in 1528, Boughton was left almost finished by Ralph (Duke of Montagu since 1705) when he died in 1709. The dukedom dying out before the end of the 18th century, Boughton passed by marriage into the possession of the Dukes of Buccleuch and since they had seats elsewhere this house was left alone without later 'improvements'. Complete with its original furnishings, it is as if preserved in amber: visiting its stately panelled and moulded rooms is like walking into a stage set for a play by Molière. Neither the gilt-and-brocade richness of the late Stuart and William-and-Mary furniture, nor the splendour of the tapestries and wall paintings soften the formality of its spaces and uncovered floors (part of which introduced the new fashion of 'Versailles' parquet to England). The finest rooms are in the state apartments. The abundance of Mortlake tapestries there is not unconnected with Lord Montagu having bought the management of the factory in 1674.

☎ Kettering (0536) 82248

3 m NE of Kettering on A43 turn E to Boughton Park

SP 9081 (OS 141)

House open Aug 1400-1700, gardens open May to Sept 1200-1800

⊖ (1 mile) P WC 🅰 🚻 D ♠ ☕ 🍴 ◆ 🐕 ♨
🏃 🐕 🧍

Castle Ashby

near Northampton, Northamptonshire

The two main periods of building activity at Castle Ashby are explicit in the entrance front: two sides of the Elizabethan house begun in 1574 are linked by a Classical screen of 1635 (possibly by Inigo Jones). Other 17th-century elements are the lettered roof balustrade with texts from Psalm 127 (begun 1624) and the reconstruction of the inside of the east side (from about 1675). The interiors are decorated and furnished mainly of that period, with magnificent carved wooden foliage, plaster ceilings (one surviving from 1620) and Mortlake tapestries. The gilded, inlaid and lacquered cabinets and the high-backed chairs are all in keeping. The Compton family, who have been the owners throughout, made alterations in the 1880s to reinstate the Jacobean-century appearance of part of the house – having previously, from the 1860s, reversed the garden scheme of Capability Brown on the north and east sides, and laid out a formal, Italianate garden with terraces. (Brown's landscaped park, with its two long grassy rides dating from the reign of William III, was left alone.) An important collection of Greek vases is on display in the long gallery (the upper part of the 1635 courtyard screen). The paintings are of high quality. There is the Antonis Mor portrait of Queen Mary I and Van Dyck's study of Charles I's favourite, the Duke of Buckingham, done after his assassination. There are fine examples by Dobson, Ramsay and Reynolds, some good Dutch pieces, and a Giovanni Bellini *Madonna* and a Mantegna *Adoration*.

☎ Yardley Hastings (060 129) 234

7½ m SE of Northampton on A428, ½ m E of Denton turn N

SP 8659 (OS 152)

Open few days each year (eg Bank Hols), phone for details; parties by appt throughout year

🅿 WC 🚻 (by appt) 🍴 🍵 (for parties, by appt) ⌐ 🚾 🦮 (compulsory) ● (not in house) ⛪

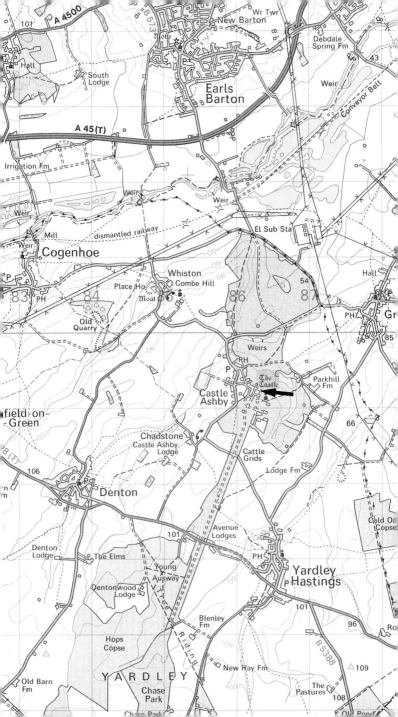

Deene Park

near Corby, Northamptonshire

Deene Park, a house that has been built over six centuries, has been the home of the Brudenell family since Sir Robert acquired it in 1514. An Elizabethan house was built here on earlier foundations, and each generation has made alterations and additions. Considerable building was done in the 18th century, so that the house has grown into a part Tudor and part Georgian mansion. The old part of the house, which includes the large block projecting towards the lake, stands round a courtyard, and the first great hall, on the left, dates from 1540. This was later turned into a billiards room, and a 13th-century doorway was found in one of the walls. The Renaissance oriel window bears the initials 'EA' for Edmund and his wife Agnes, whose fortune paid for the building of another and larger great hall in 1571. The main front, facing across the lake, has a central part of about 1810 in a simple Gothick style. The visitor enters the house through a door in the east range, the old part of the house, forming the old hall. This once rose through three storeys to an open roof, but when Sir Edmund built the new great hall he divided it horizontally into three rooms. Sir Edmund's hall has some fine panelling and a complex hammerbeam roof as well as a refectory table and matching seat of about 1560. The stained glass was put in in the 17th century by Sir Thomas Brudenell, later 1st Earl of Cardigan. The dining room has portraits of the horses of Lord Cardigan, hero of Balaclava, and a painting showing him leading the Charge of the Light Brigade.

☎ Bulwick (078 085) 278/361/223

7 m NE of Corby on A43 turn W to Deene

SP 9592 (OS 141)

Open Jun to Aug Su also Bank Hols Easter to Aug 1400-1700; by appt at other times

⊖ P WC ⬚ ⊟ ♣ ⬤ ⊓ ◆ ⚹ ⚐ (by appt)
⬤ (not in house)

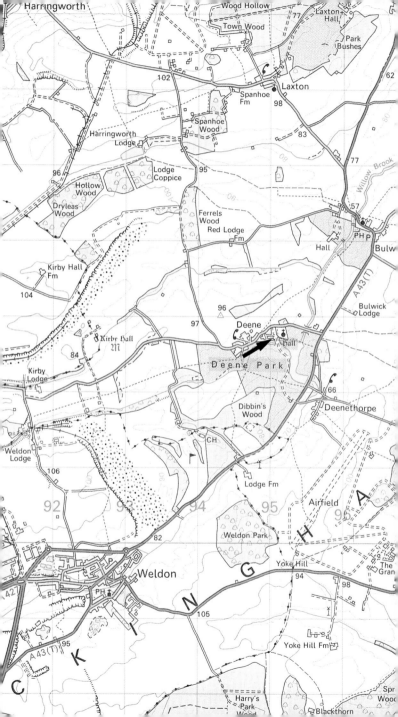

Kirby Hall

Deene, near Corby, Northamptonshire

Kirby Hall is an uninhabited ruin, lacking the greater part of its roof, but it is still one of the most beautiful Elizabethan buildings in the country. It was begun in 1570 by Sir Humphrey Stafford and completed after his death by Sir Christopher Hatton, who bought the property in 1575. Important alterations were made in 1638-40 for Christopher Hatton III, and the Hattons owned and occupied the Hall until 1764. It then passed down the female line to the Finch-Hattons, whose descendants still retain the estate. The buildings are arranged round a central courtyard, and the visitor enters through the porch in the north range, which was almost certainly altered by Inigo Jones. The main south range containing the great hall is opposite, while the ranges at the sides contained the 'lodgings' or guest rooms. The courtyard itself is essentially Sir Humphrey Stafford's building, and all four elevations are decorated with Giant Orders of pilasters which ascend the whole height of the building, the first known use of such a device in England. The great hall is entered through the lovely decorated two-storey porch, French in inspiration, and dated 1572 at the base of the gable (the carved plaster ceiling was added in the 17th century). The hall still has its original timber roof, with corbels decorated with carved devices of the Staffords, while the gallery dates from about 1660. The great staircase was built by Christopher Hatton, as was the south-west wing, the two main rooms of which are lit by great bay windows, among the earliest known in the country.

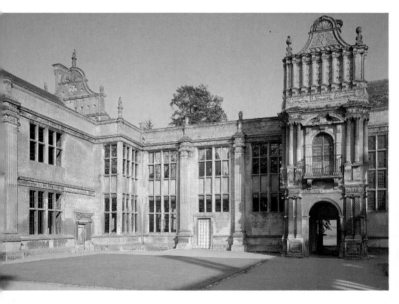

☎ Corby (053 63) 3230

7 m NE of Corby on A43 turn W to Deene, continue for 3 m then turn N

SP 9292 (OS 141)

Open daily throughout year, mid Mar to mid Oct 0930-1830, mid Oct to mid Mar 0930-1600

♿ WC ♿ (limited access) 🚻 D (on lead, grounds only) ♠ ◆ ✄ EH

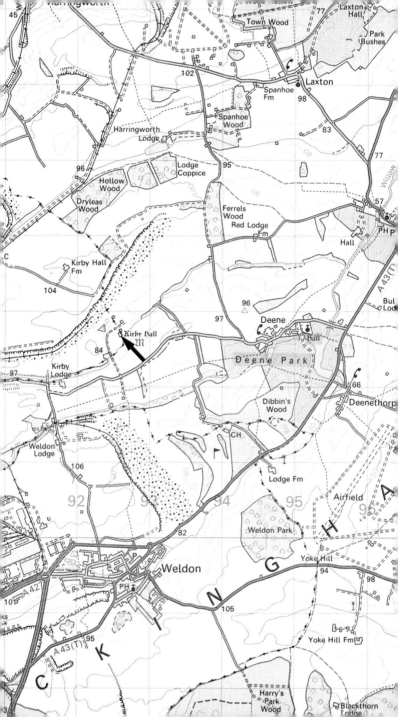

Lamport Hall

Lamport, Northampton, Northamptonshire

The Isham family held the manor of Isham from the Norman Conquest, and possibly before, but although there was rebuilding in the Elizabethan era, very little now survives. The oldest part of the house is the 'Italian palace' of 1655, built for Sir Justinian Isham by John Webb. This now forms the centre of the main front; the wings were built by Francis Smith of Warwick between 1732 and 1738, improvements being made to the interior at the same time. A new staircase was built, and much of Webb's Classical decoration in the high room, now called the music room, was replaced with exuberant ornamental plasterwork. Webb's fine chimneypiece, however, carved by Cibber with the Isham swans, survives. Further alterations and rebuilding took place in the first part of the 19th century, and the north entrance front was built for Sir Charles Isham in 1861 to designs of William Burn. The house contains fine furnishings, including two large 17th-century cabinets with mythological scenes painted on glass and some excellent 17th- and 18th-century Chinese porcelain. An unusually complete collection of family portraits gives an unbroken record of 400 years of Ishams, and there is a huge painting of Queen Anne of Denmark, who visited Pytchley Hall, residence of another branch of the family, in 1605. The gardens were the main interest of Sir Charles Isham, who appears to have invented the garden gnome – his alpine rockery was once alive with tiny figures, all made in Nuremburg. Only one of these now survives, and is shown in the hall.

☎ Maidwell (060 128) 272

10 m N of Northampton on A508

SP 7574 (OS 141)

Open Easter to end Sept Su and Bank Hol M also Th in July and Aug 1415-1715

⊖ P WC ⊟ (by appt) D (grounds only) ♣ ☛
⊓ ◆ ♨ ⚹ (by appt)

Southwick Hall

near Oundle, Northamptonshire

Southwick Hall, a manor house built over the centuries, has been the home of three inter-related families, the Knyvetts, the Lynns and the Caprons, who still live here today. The Knyvetts built the medieval house, parts of which still remain, notably the two towers. The Lynns, who owned the manor from 1441 to 1840, rebuilt in Tudor times and again in the 18th century, when George Lynn, a cultured man with antiquarian and scientific interests, extended the house on the west side and 'Georgianised' a large part of the interior. The Caprons, who bought the house in 1841, later rebuilt the east wing, built the stable block, and in 1909 altered the entrance through the undercroft. The interiors of the main rooms, particularly the hall, clearly illustrate the three main periods of building. The hall, originally the great hall built about 1300 by John Knyvett, was rebuilt in 1571 to become a ground-floor room with two bedrooms above; in the 18th century it was stripped of its wainscotting and given Georgian details. The oldest part of the house is the vaulted room and circular stair turret built about 1300, while the undercroft, or crypt, was part of the extension built about twenty years later by John's son Richard, as was the Gothic room, which was apparently used as both a living room and a chapel. The middle room, once the solar, was rebuilt by George Lynn in 1580; the Elizabethan windows still remain, and one of the Elizabethan bedrooms over the hall has its original panelling. The bed, made in about 1640, is still in use.

☎ Oundle (0832) 74013/74064

8 m NE of Corby on A43 turn E just before Bullwick

TL 0292 (OS 141)

Open Apr to end Aug Bank Hol Su and M also w 1430-1700; parties by appt at other times

P WC 🅰 (limited access) 🚽 D (grounds only)
♣ 🍴 ⛱ ♨ 🏃

Holme Pierrepont Hall

Nottingham, Nottinghamshire

Holme Pierrepont, one of the earliest brick buildings in the county, and originally a courtyard house, shows the work of three centuries. The brickwork of the Tudor front range dates from about 1490, and behind this survives intact a timber roof whose beams run the whole length of the house. Considerable enlargements were made in the 17th century, but apart from one range, most of this Jacobean work was demolished in the 1730s, and a Victorian range was added in the 1870s. The three ranges are grouped round a courtyard garden, which has been laid out by the present owners to the original plan of 1875, with a formal box parterre, rose beds and a huge herbaceous border. Visitors are free to walk through the house at will. The grand staircase, which has elaborately carved floral panels, probably came from the demolished Jacobean hall range, as it dates from the time of Charles II; and the 'big room' has a lovely plaster ceiling of the 1660s recently brought from another house, and 18th-century walnut furniture. Some of the other rooms have Tudor fireplaces and doorways, and all are lived in by the family, who are carrying out extensive restoration work. The early Tudor 'lodgings' range, which still has its original garderobes (lavatories), has been fully restored, with the ornamental timber roof beams exposed and the rooms well furnished with pieces collected by the family over the last 250 years. Holme Pierrepont is also the setting for the authentic productions of Baroque operas staged by the Holme Pierrepont Opera Trust.

☎ Radcliffe-on-Trent (06073) 2371

2 m S of Nottingham on A52 take 1st turning E after River Trent to Adbolton

SK 6239 (OS129)

Open Easter and late May Bank Hol weekends Su-T; May Day M, June to Aug Su, T, Th, F 1400-1800

♿ (1 m) 🅿 WC 🚻 (limited access) 🚌 (by appt) ♣ 🍴 ◆ ⚘ 🐕 (parties only) ●

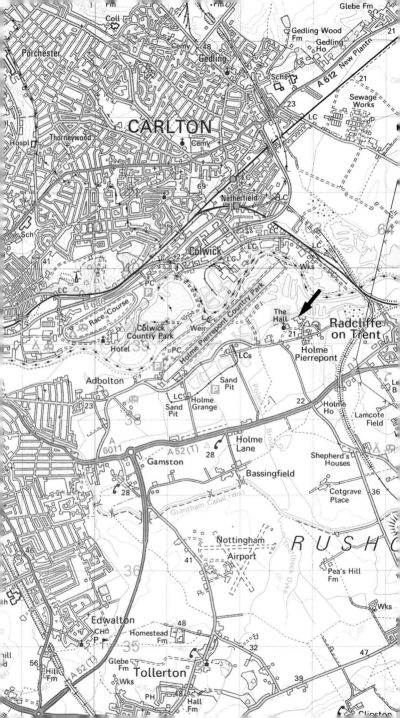

Blenheim Palace

Woodstock, Oxfordshire

In gratitude for the Duke of Marlborough's victory at Blenheim in 1704, Queen Anne grantd him the Royal Manor of Woodstock and money to build a hero's mansion. Sir John Vanbrugh, assisted by Nicholas Hawksmoor, designed it as a celebration of the Duke's triumphs; he succeeded in creating the greatest Baroque mansion in England. The size of the main façade, with the huge stable and service blocks set back on either side of the main court, is overwhelming. The scale is maintained in the towering portico over the entrance and in the hall, where the painting on the ceiling by Sir John Thornhill commemorates the battle. In the saloon the ceiling and walls were painted by Louis Laguerre. On the west side the gallery, 180 foot long, has a bay window looking out over terraced gardens to the lake. Among the many treasures three paintings in the family apartment summarise its history: the portrait of the 1st Duke by Kneller; Sir Joshua Reynolds' portrait of the 4th Duke with his wife and six children; and Sargent's portrait of the 9th Duke with the beautiful Consuelo Vanderbilt and their two sons. The 9th Duke did much to restore the palace, after a long period of neglect, early this century. The gardens and park were included in Vanbrugh's design. The gardens were laid out by Queen Anne's gardener, Henry Wise, while building was in progress, but the park remained unfinished until the 4th Duke employed Capability Brown to complete the landscaping in 1784; many consider it his greatest achievement.

☎ Woodstock (0993) 811325

11 m NW of Oxford on A34

SP 4416 (OS 164)

Palace open mid Mar to end Oct daily 1100-1800; park open throughout year 0900-1700

♿ 🅿 WC ♿ (preferably by appt) 🚻 D (on lead, grounds only) ♣ 🍴 ⛲ ◆ 🎎 ⚘ 🏌 ● ⛺ ☂

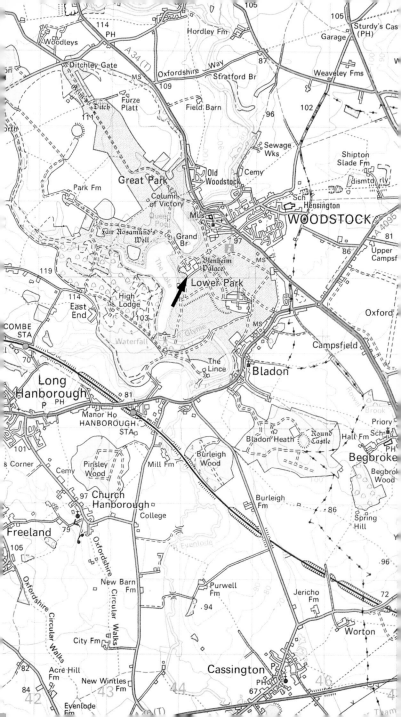

Chastleton House

near Moreton-in-Marsh, Oxfordshire

Chastleton has been scarcely touched by changing fashions – it looks now as it did in the opening years of the 17th century, when it was built for Walter Jones, a successful wool merchant. The architect is not known – it may have been mainly Jones himself – but the front, of golden Cotswold stone, is a sophisticated design, with projections and recessions, windows set at different levels, and towers containing staircases on each side. The house is built round a courtyard, and the front door, which is set sideways, opens into the screens passage and thence into the great hall. This hall has an ornately carved screen and a long oak refectory table. The inside of the house is complex and bewildering, with a great many rooms, but the finest is the great chamber on the first floor, with its painted and carved panelling, vast ornate chimneypiece bearing the arms of the Joneses, and equally ornate plaster ceiling. Several of the other rooms also have plaster ceilings and good fireplaces, and much of the furniture in the house, like the refectory table in the hall, has been there since the house was built. Later additions, such as early lacquer cabinets, blend well, and there are some fine tapestries and hangings, including crewel-work made in the house. Both staircases, east and west, lead up to the great tunnel of the long gallery, which runs right across the top of the house and has a splendid barrel-vaulted ceiling decorated with strapwork and flowers.

☎ Barton-on-the-Heath (060 874) 355

5½ m NW of Chipping Norton on A44 turn SW

SP 2429 (OS 163)

Open Good Fri to last Su Sept F, S, Su and Bank Hol M 1400-1700

🅿 WC ♿ (at other times by appt) D (on lead, grounds only) ♣ ✿ 🛈

Kingston House

Kingston Bagpuize, Oxfordshire

There is some doubt about the date of this compact red-brick manor house, nor is the architect known. Its Baroque feeling, and its similarity to some of the work of Wren and Gibbs, have led some experts to claim that it was built about 1710, but the family have found deeds that show it was in existence by 1670, when John Latton sold it to Edmund Fettiplace. Tours, which are conducted by members of the family, start in the staircase hall. The staircase, magnificently cantilevered and with no supporting pillars, is one of the important features of the house. The saloon, which was originally the entrance to the house, is a good vantage point from which to appreciate the lovely proportions of the house, with its symmetry of design, high rooms and fine architectural detail. The rooms are all of a similar character, with Queen Anne fireplaces and panelling in some cases. There are some good paintings, though no major works, and the furniture is mainly 18th century, a mixture of French and English. The library contains a longcase clock showing the phases of the moon and has an intricately carved chimneypiece in the style of Grinling Gibbons. The charming small morning room, which may have been a bedroom, has panelling of an earlier type than the other rooms. Some of this was removed fifty years ago and found in the stables by Miss Raphael, the previous owner, who replaced it. Miss Raphael also laid out the English garden, which is kept up by her niece Lady Tweedsmuir, the present owner.

☎ Longworth (0865) 820259
5½ m W of Abingdon on A415 at Kingston Bagpuize
SU 4097 (OS 164)

Open May, June and Sept W, Su and Bank Hol M, also Aug Bank Hol M 1430-1730; parties by appt at other times; no children under 5 in house

⊖ 🅿 WC 🚻 (by appt) ♣ 🍴 ◆ ⁂ 🐾 ●

Kingstone Lisle Park

near Wantage, Oxfordshire

This fine house with its Classical proportions is very much in the style of Sir John Soane, and owes its character to the extensive alterations made in about 1812 by either Cockerell or Basevi, both of whom had been apprenticed to Soane. The two wings were added at this date, though the central part of the house dates from 1677, and one of the wings, that housing the billiards room, has dummy windows on the approach side, making the house appear larger than it actually is. The interior is noted particularly for its superb 'flying staircase', one of the finest in the country, though the designer is not known. The drawing room contains some good late-18th-century furniture and two unusual carpets, modern, though in traditional Morris designs, and made by hand, using the embroidery techniques of gros- and petit-point. There is much exceptional needlework throughout the house, and the card table, stools and firescreen were all worked by the mother of Mrs Lonsdale, the present owner. The room also contains a collection of glass made by Captain Lonsdale, several pieces of miniature furniture and an interesting portrait by Marc Gheeraerts of Elizabeth Throckmorton, the wife of Sir Walter Raleigh. The morning room has several good paintings, including an unusual view of Whitehall by Hogarth, a small Constable and a Norfolk landscape by John Crome. The furniture here is mainly Queen Anne, and the tapestries were worked by Mrs Lonsdale, as was the splendid carpet in the sitting room, which was designed for the room and took seven years to complete.

☎ Uffington (036 782) 223

W of Wantage on B4507

SU 3287 (OS 174)

Open Apr to Aug Th and Bank Hol weekends 1400-1700; parties at other times by appt

🅿 WC ♿(limited access) 🚻 D (on lead) ♣ ☛
★ ✿ ⚔ (by appt) ● (not in house)

Mapledurham House

near Reading, Oxfordshire

The house, an attractive Elizabethan building of patterned red brick, was built between 1581 and 1612 by Sir Michael Blount, whose family have lived here ever since. Built on an H-plan, with the central range containing the main rooms, it was altered in the 18th century, when battlements were added to replace the old roof gables. At the same time a small chapel was built in the 'Gothick' style, the great hall was divided up and also 'Gothicked', and sash windows were put in instead of the original mullions. In the 1830s, however, the front was changed back again to more or less its original appearance and a porch was added. The tour of the house starts in the entrance hall, part of the 18th-century alterations. The bleached oak panelling, from another house, was installed in 1863, and there is an interesting and unusual collection of 17th- and 18th-century carved wooden animal heads – such as the wolf in sheep's clothing – which apparently symbolise virtues and vices. The elaborate main staircase is the original Tudor work, and there is still some 17th-century plaster decoration to be seen. The house contains good collections of 17th- and 18th-century furniture and paintings, including a portrait by Sir Godfrey Kneller of Alexander Pope. The latter came frequently to the house between 1707 and 1715 to visit Martha and Teresa Blount, and when he died in 1744 he left his possessions to Martha. The state bedroom has a fine four-poster bed dated 1760, covered with a marvellous 19th-century patchwork quilt containing 15,700 pieces.

☎ Reading (0734) 723350
4 m NW of Reading on A4074 turn SW to Trench Green
SU 6776 (OS 175)

Open Easter to end Sept S, Su and Bank Hols 1430-1700; parties only T, W, Th pm
WC 🚻 ♿ 🍴 ⛱ ◆ ✻ ✗ (parties by appt)
● (not in house)

Milton Manor

Abingdon, Oxfordshire

This graceful red-brick house, built in 1663 for Thomas Calton, presents something of a mystery, as tradition says it was designed by Inigo Jones, and it is referred to by Bryant Barrett, a later owner, as 'the Manor House built on a design of Inigo Jones'. Jones had, in fact, been dead for eleven years, but Barrett's words are probably quite literal – Jones's designs continued to be used for some time, and in any case his ideas were still much in vogue. The house, a simple square three storeys high, has an unusual feature in that the front and back are identical, and all four roof elevations are precisely equal. In 1764, Bryant Barrett, a London lacemaker and devout convert to Roman Catholicism, bought the house (his descendants live here still), and with the help of his architect Stephen Wright, added the two Georgian wings and all the outhouses, including stables, a bakery and a brewery. He also landscaped the garden, keeping detailed notes of all his plantings. The new wings, quite plain outside, are far from plain inside. The hall has a marvellous 17th-century carved fireplace and a 16th-century monk's settle said to have belonged to the abbots of Abingdon; the charming drawing room still has its original carved ceiling, and most of the rooms have fine Georgian panelling. But the high spot is undoubtedly the library in the Strawberry Hill 'Gothick' style, on which the well-known carver Richard Lawrence worked with Stephen Wright. The room contains a portrait of Barrett and his family, and a telescope belonging to Admiral Benbow.

☎ Abingdon (0235) 831287/831871
16 m N of Newbury on A34 turn NE at junction with A4130 to Milton
SU 4892 (OS 164)

Open Easter to end Oct S, Su, Bank Hols 1400-1730; parties at other times by appt

⊖ P WC ☒ ☐ (appt only) D (grounds only) ♣ ♥ ☂ ☚ ⚔ (compulsory) ● (not in house)

184

Rousham Park

Rousham, Steeple Aston, Oxfordshire

Rousham, together with its garden, is one of the finest examples of the work of the great 18th-century architect William Kent, and the garden, one of the first landscape designs in the country, is almost as he left it. The house itself, built around 1635 for Sir Robert Dormer, was on the common Tudor H-plan with mullioned windows and gables. Kent, who was called in by General James Dormer in 1738 to make improvements to both house and garden, turned it into a sort of Tudor palace with Gothic overtones. He gave it a straight battlemented parapet instead of the gables, glazed the windows with octagonal panes and built two low wings, that on the garden side having Gothic ogee niches containing Classical statuary. The south front has been little altered, but unfortunately Kent's windows were replaced with sash windows. Kent improved most of the rooms by inserting new fireplaces and doors, but left untouched the original Jacobean staircase and one small room, which retains its 17th-century panelling. Two of the rooms he created are particularly memorable, the painted parlour and the great parlour. The former is purely Palladian, with a marvellous marble chimneypiece, richly carved, as is every other architectural detail. Kent supervised every detail, designed the furniture, and painted the mythological scene on the ceiling himself. The great parlour, more romantic in feeling, has an extraordinary ribbed and vaulted ceiling. It was originally the library, but in 1764 the bookshelves were replaced by the Rococo frames that now adorn the walls.

☎ Steeple Aston (0869) 47110

14 m N of Oxford on A423 turn NE to Rousham

SP 4724 (OS 164)

House open Apr to Sept W, Su and Bank Hols 1400-1630; gardens open daily. Parties at other times

♿ (exc Su) 🅿 WC 🚻 (by appt) ♠ 🛋 ✗ 🗡 ●

No children under 15 admitted

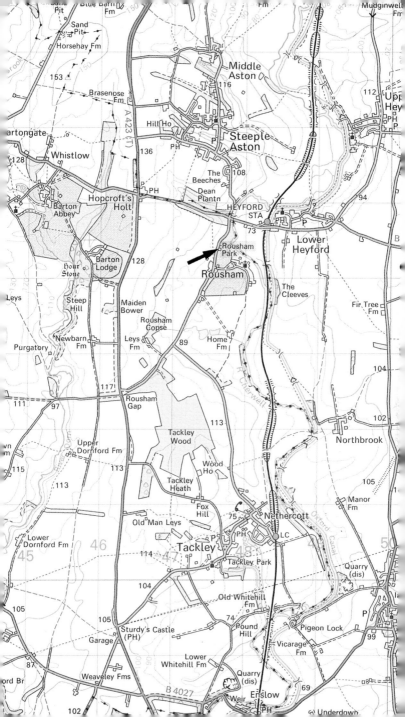

Stonor

Henley-on-Thames, Oxfordshire

The house, hidden in a fold of the Chilterns and surrounded by beechwoods, is largely 18th-century in appearance, but its long, low, red-brick Georgian façade fronts the greater part of a large medieval house. The Stonor family, who have always been Roman Catholics, have lived here since the 12th century, and the heavy penalties they had to pay after the Reformation for their beliefs prevented them from doing much new building. The house is still, in its core, a group of medieval buildings, the first recorded ones being the old hall, buttery, solar and flint-and-stone chapel all built between 1280 and 1331. For the next two hundred years there was constant building and renovation, and in the 16th century Sir Walter Stonor joined the buildings to make a formal E-shaped Tudor house and built a many-gabled brick front. The house was altered again in the 1750s by the architect John Aitkins, who removed all but the central gable, put in sash windows and added the present roof with its heavy cornice. He also redecorated the hall in the 'Gothick' style – the same style was later used for the chapel. There are some attractive and well furnished rooms, and several have been redecorated. The dining room has early-19th-century wallpaper showing the buildings of Paris; there is a bedroom with an 18th-century French bed and chairs shaped like seashells, and the house contains important Renaissance bronzes and good tapestries. The library holds a large collection of books relating to secret Roman Catholicism.

☎ Turville Heath (049 163) 587
6 m N of Henley-on-Thames on B480 turn E after Stonor
SU 7489 (OS 175)

Open Apr Su pm, May to Sept W, Th, S (Aug), Su 1400-1730; Bank Hol M (inc Easter) 1100-1730

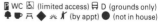 P WC 🚻 (limited access) 🚌 D (grounds only) ♣ 🍽 🎪 ◆ ♨ 🏹 (by appt) ● (not in house)

Sulgrave Manor

Sulgrave, near Banbury, Oxfordshire

The fame of this charming 16th-century manor house rests on its historical associations, as it was the home of the ancestors of George Washington. His great grandfather, John, emigrated in 1657. The land came into the Washington family when Lawrence Washington bought it in 1539 – for the sum of £324 14s 10d. Lawrence was at the time Lord Mayor of Northampton and reasonably prosperous. He also had four sons and seven daughters, which presumably necessitated a larger dwelling. The house he built, a modest one of the local stone, was completed in 1560, but there have been several alterations, and it passed through many hands and became quite dilapidated before being purchased by subscription in 1914 and restored as a memorial to Anglo-American friendship. The most valuable item is a fine portrait of George Washington by the famous American painter Gilbert Stuart (1755-1828), but there are many other relics of the great man, and two small rooms are used as a Washington museum. The manor has been very carefully restored, and there are some fine fittings and pieces of furniture, including an Elizabethan four-poster bed with an embroidered coverlet from the reign of William and Mary. During work on the great hall an Elizabethan sixpence and a baby shoe were found, together with a knife case which evidently belonged to Lawrence Washington himself. The kitchen is particularly interesting as it contains a 200-year-old kitchen bought, complete with all its fittings, from a manor house in Hampshire.

☎ Sulgrave (029 576) 205

8 m NE of Banbury on B4525 turn E after Thorpe Mandeville to Sulgrave

SP 5645 (OS 152)

Open Feb to end Dec daily exc W; Apr to end Sept 1030-1300, 1400-1730; Oct to end Mar closes 1600

P WC ⛟ ♣ ☕ ◆ ⁂ ✗ ● (not in house) School groups welcome

Chillington Hall

near Wolverhampton, Staffordshire

Chillington, home of the Giffard family, has had one of the longest family successions in the country, having passed down directly from father to son since 1178. The house itself, however, is almost entirely 18th century, though a Tudor house certainly existed, and the plan of the 18th-century building was dictated by the original layout. The south side was rebuilt for Peter Giffard in 1724, but the main front is part of a major building campaign carried out in the 1780s by Thomas Giffard. The latter, fresh from his obligatory travels in Italy, engaged the Classical architect John Soane to rebuild the house, and it is an interesting example of Soane's earlier work. The remaining Tudor buildings were demolished, but Soane was forced to retain the 1724 range, and had to alter his design to accommodate them. The columned entrance hall and the three tall ground-floor rooms with their huge windows show the fine quality of his work. The saloon, which is built within the Tudor great hall, is more typical of his work: it is lit by means of a domed roof, an experiment in overhead lighting that he was to develop in the Bank of England. A small drawing room is shown in the largely private 1724 range, and the fine oak staircase with Italian-style plasterwork and the family's crest carved into each bracket. This leads to an elegant arched and vaulted corridor by Soane, and thence to the state bedroom, where there is a splendid painted bed. The park, with its huge lake and ornamental bridges, is one of the finest examples of Capability Brown's work.

☎ Brewood (0902) 850236
6½ m N of Wolverhampton on A449 turn NW to Brewood for 3 m then SW
SJ 8606 (OS 127)

Open May to mid Sept Th, Su of Bank Hol weekends and Su in Aug 1430-1730

🅿 WC ♿ (limited access) 🚻 D ♣ ⚹ ✗ ● ⚘

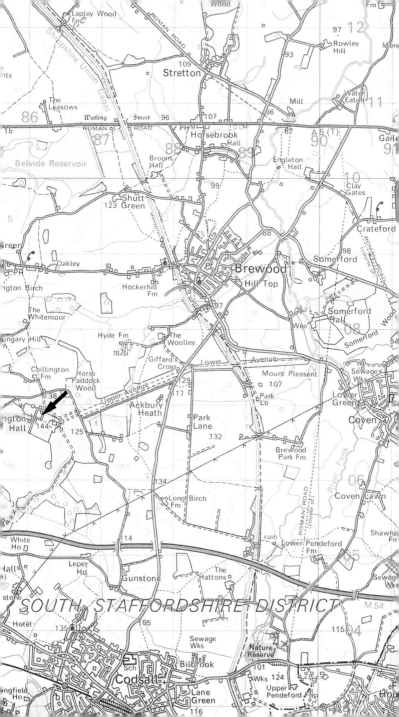

Hanch Hall

Lichfield, Staffordshire

Hanch Hall has associations going back to the 13th century, when the Aston family built a dwelling here, but very little is known about the early history of the house. There was clearly some building in Tudor times – 16th-century timbers can be seen in one room, and the cellars are Elizabethan, but little else remains. The fine red-brick and stone range facing the garden is 18th century, and the rest of the exterior seems to date from around the early Victorian period. The guided tours, which take in eighteen rooms, are conducted by members of the family, who bought the house in 1975 and are restoring it themselves. One of the best features is a splendid early Victorian oak staircase with carved decoration in the 'Elizabethan' style, and the windows above it have armorial glass recording the families who have lived here. Most of the downstairs rooms have old panelling, and there is some good 18th-century furniture, including a fine Georgian four-poster bed supposed to have been used by the poet Shelley. There are also some *boulle* and marquetry pieces, and the 19th-century ballroom contains a stool of the 1920s made of stuffed leopards' feet. The upstairs rooms, which are arranged to illustrate different periods, display various collections: dolls, needlework, porcelain, teapots, costume, christening gowns and even one called 'Postmen through the Ages'. The grounds, which are open to visitors, contain free-roaming peacocks, pheasants and other wildfowl.

☎ Armitage (0543) 490308

4 m N of Lichfield on A515, turn N onto B5014 for ¾ m

SK 1013 (OS 128)

Open Apr and May Su, Bank Hol M and following T; June to Sept T-Th, S, Su 1400-1800

🅿 WC 🚻 (limited access) 🚗 ♣ 🍴 ◆ ☸ ● (not in house) ⚑ (candlelit tours by appt)

Hoar Cross Hall

Hoar Cross, near Burton upon Trent, Staffordshire

This huge seventy-roomed Elizabethan-style mansion, set in twenty acres of woodland and gardens, was built at the height of Victorian prosperity for Hugo and Emily Meynell, to replace the old Elizabethan house which stands nearby. Hugo died in a hunting accident in 1871, the year the house was completed, and his widow built the adjoining Church of the Holy Angels in his memory. The architect of this fine Victorian building was G. F. Bradley, who was also responsible for some of the interior work in the house. The house itself was designed by Henry Clutton, and is large and imposing if not adventurous. The interiors are safely Elizabethan in style, with oak panelling and richly moulded plaster ceilings. That in the long gallery is the work of Bodley, and so is the fine carved screen in the hall, standing astride a short flight of steps. The best of Bodley's work, however, is the chapel, built in 1897, with a richly decorated ceiling, much gilding and carving, stained glass, and panelled walls in the Henry VII style. Between the Second World War and 1970, when the property finally passed out of the Meynell family, the house and gardens fell into a state of some decay. Much restoration work has been done on the house since then, and the grounds are now also being restored. There are some fine and rare trees, and a pair of wrought-iron gates attributed to the famous Robert Bakewell of Derby. Hoar Cross Hall is well known for its medieval banquets, which are held throughout the season.

☎ Hoar Cross (028 375) 224
12 m W of Burton upon Trent on B5234, turn S to Hoar Cross
SK 1223 (OS 128)

Open end May to early Sept Su and Bank Hol M, also at other times, telephone for details
Ⓟ WC ♿ (by appt) ⊟ ♣ ⬤ ⊓ ◆ ☼
⬤ (not in house) 🏃 (parties by appt) ☂

Shugborough

Milford, near Stafford, Staffordshire

This elegant house, set in its lovely 18th-century park, has been the home of the Anson family since 1624. The three-storey central block was built for William Anson in 1693, and in the 1740s Thomas Anson added the pavilions on either side, with their domed bow windows. He later commissioned the architect 'Athenian' Stuart to make further alterations and to design the buildings in the park. These monuments, temples and pavilions which are superb examples of the Greek Revival style and are more famous than the house itself, were financed by Thomas' younger brother, Admiral Lord Anson. The last important additions were made between 1790 and 1806 and the architect was Samuel Wyatt who added the huge colonnaded portico on the main front. He also cased the whole exterior in slates painted to resemble stone, but most of these were later replaced with stucco. The oval shape of the hall was Wyatt's creation, and he introduced a ring of *scagliola* (imitation marble) columns with Doric capitals based on those found at Delos. The plasterwork frieze by Joseph Rose incorporates the Anson crest. Most of the rest of the interior is Wyatt's work, though the library and dining room have kept their original 1740s decoration. The most impressive room is the red drawing room. The coved ceiling with its Neo-Classical ornament is one of the finest works of the famous plasterer Joseph Rose, and there is an elegant white marble chimneypiece with ormolu mounts. The house contains some excellent French furniture, family portraits and sporting paintings.

☎ Little Haywood (0889) 881388
6 m SE of Stafford on A513, turn NE to Shugborough Park
SJ 9922 (OS 127)

Open Mar to Oct T-F 1030-1730, S, Su 1400-1730; Nov to Feb T-F 1030-1630, Su 1400-1640

⊖ (1 m) 🅿 WC 🖐 (limited access) 🚻 D (gardens only) ⚶ ⚔ (parties by appt) ● NT

Whitmore Hall

near Newcastle-under-Lyme, Staffordshire

Although no more than three or four miles from the Potteries, Whitmore Hall is charmingly manorial and rural in setting. It forms a unit with the estate village and the little medieval church, and is shielded from the industrial scene by the woods of Swynnerton Old Park. The visitor passes down a broad avenue of limes from the churchyard to the south front of rich red brickwork with stone dressings and a stone balustrade at the top; behind the façade, the house is actually a four-storey structure. The entrance hall, with its twin Corinthian columns, dates from the Georgian period, and the porch is mid 19th-century. However, within its 17th-century casing of brick lies a much older, timber-framed house. The history of the estate can be traced back to John de Whitmore, who held it in the late 12th century. Over the centuries it was always passed down by inheritance. Towards the end of the 14th century Elizabeth de Whitmore brought it to the de Boghays by marriage. Then in 1546 Alice de Boghay married Edward Mainwearing; and there have been Mainwearings at Whitmore ever since. The stables at Whitmore are particularly admired, since they provide an unusual example of early 17th-century craftsmanship in wood; the interior fitments – including the horses' stalls, with their turned columns and arches – are wonderfully well preserved. Among the other attractions of Whitmore are the elegantly landscaped park and lake behind the house.

☎ Whitmore (0782) 680235

4 m SW of Newcastle-under-Lyme on A53

SJ 8141 (OS 118)

Open May to Aug T and W 1400-1700; parties at other times by appt

⊖ 🅿 WC ☐ (by appt) ♠ ✲ 🏃

Arbury Hall

Nuneaton, Warwickshire

One of the best examples in the country of the 18th-century 'Gothick' style, Arbury, originally an Elizabethan courtyard house, was largely the creation of Sir Roger Newdigate from 1748, using a variety of architects. Mary Ann Evans, later known as George Eliot, was born on the estate, and described it in one of her early books, writing of the 'transformation from plain brick into the model of a Gothic manor-house'. In the 1670s Sir Richard Newdigate had built the impressive stables (partly to a design by Wren) and employed a plasterer who had worked on Wren's City churches to redecorate the chapel; but the character of the house comes from the 18th-century work. The 'Gothicking' continued gradually up to Sir Roger's death in 1806, the earlier parts being contemporary with Horace Walpole's Strawberry Hill, and the architects included Sanderson Miller, Henry Keene and Thomas Crouchman. Miller also advised on the gardens, which were landscaped by Sir Roger. The style of the house is a light and exuberant version of the true Perpendicular Gothic, and the lovely, delicate vaults are not of stone but of plaster. The tour covers most of the finest rooms, and it is interesting to see the changes in style as the work progressed. The drawing room, of 1762, has a very pretty plaster ceiling, while the dining room, designed by Keene ten years later, has a fine fan vault, but the plasterwork vaulting really reaches its height in the spectacular saloon ceiling, with its pendant drops and lace-like tracery; the inspiration was Henry VII's chapel in Westminster Abbey.

☎ Nuneaton (0203) 382804

2½ m S of Nuneaton on A444, turn W to Arbury Park

SP 3389 (US 140)

Open Easter to end Sept Su, Bank Hols also T and W in July and Aug 1400 1700

 (1 m) ⓟ WC ◩ ⊟ (by appt) D (grounds only) ♣ ⬥ ◆ ⊀ (compulsory) ●

Charlecote Park

Wellesbourne, Warwickshire

Charlecote was built in the 1550s for Sir Thomas Lucy. The house belonged to the Lucy family from the 12th century until 1948, but only the gatehouse survives intact from the Elizabethan building, as George and Mary Elizabeth Lucy made extensive alterations in the 1820s and 1850s. Their avowed intention was to restore the house to its former Elizabethan splendour rather than to alter it, and they made it their life's work, scouring the Continent for suitable fireplaces, paintings and marble pavings, and using only the very best in craftsmen and materials. Thomas Willement, a fashionable designer best known for his stained glass, advised on the decoration, and John Gibson, a pupil of Barry, was employed to design the church as well as some of the interiors. Inside the house there is no hint of the building Elizabeth I would have seen; the rooms are all 19th century and all arranged to look just as they would have done in the 1860s. The richly coloured decorations in the library, dining room and great hall are the work of Willement, while John Gibson's hand can be seen in the ebony bedroom and the drawing room in the north wing, dating from the 1850s. The hall contains an interesting collection of family portraits as well as a huge table bought from Fonthill Abbey in 1823 by George, who had much admired William Beckford. More of the Fonthill furniture can be seen in the tapestry bedroom. The spectacular carved and ornamented sideboard in the dining room, made by George Willcox in 1858, was bought by Mrs Lucy for £2000.

☎ Stratford-upon-Avon (0789) 840277

5 m E of Stratford on B4086

SP 2656 (OS 151)

Open Apr to end Sept daily exc M and Th but inc Bank Hol M 1100-1800 (or dusk if earlier)

P WC ♿ ☐ (by appt) ♠ ▭ ⩎ ◆ ⚘ ● (no flash or tripods) NT

Lord Leycester Hospital

Warwick, Warwickshire

The hospital is formed of a group of buildings, the oldest of which is the Chapel of St James, built over the archway of the town's west gate in the late 14th century. The chapel became the property of the Guild of St George, which was granted its charter in 1383 and later merged with two other guilds to become the United Guilds of Warwick, and the great hall was built by them soon after this (it still has its original timber ceiling). This was the United Guilds' public room, for such functions as assemblies and feasts, while the Guildhall, built by Neville, Earl of Warwick, about 1450, was the private meeting chamber, where the public was not admitted. The hospital itself was founded in 1571 as a haven for aged and infirm retainers and their wives by Robert Dudley, Earl of Leicester (Leycester), and the Guildhall was divided up into living quarters which, although fairly primitive, continued in use until 1956. The building was then found to be in a dangerous condition, and the original Corporation of the Master and Brethren, which had been in existence since the charter from Elizabeth I, was replaced by a Board of Governors. The extensive restoration work was completed in 1966; eight ex-servicemen and their wives now live here in modernised flats and help with the running of the hospital. The buildings, in spite of some necessary alterations, retain a genuinely medieval feeling: the Master's House has been in continuous occupation since 1400, and the delightful courtyard bears the arms and devices of Lord Leycester and the Sidney family.

☎ Warwick (0926) 492747/491422

In centre of Warwick by the west gate

SP 2865 (OS 151)

Open daily exc Su, Good Fri and 25 Dec; summer 1000-1730, winter 1000-1600

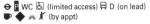
♿ 🅿 WC ♿ (limited access) 🚽 D (on lead)
🍴 ◆ ※ 🏃 (by appt)

Ragley Hall

Alcester, Warwickshire

This large and imposing house, in its lovely park laid out by Capability Brown, was built between 1679 and 1683 for Sir Edward Seymour, ancestor of the Marquess of Hertford, whose home it is today. The original architect was the scientist Robert Hooke, contemporary of Wren, and it is the only surviving example of his building work. In 1750 James Gibbs made improvements to the interior, notably the fine Baroque entrance hall with Rococo plasterwork by Atari, and the stately portico was added by James Wyatt, who also redesigned several of the state rooms in anticipation of a visit from George III. One of his rooms is the red saloon, with delicate Adam-style plasterwork and walls hung with crimson silk, and it is still almost exactly as he left it. Wyatt also designed the stable block, which still houses horses and carriages. During the Second World War the house was used as a hospital, and thereafter became almost derelict, but many of the rooms have since been redecorated in colours similar to the original ones, and the owners are still in the process of restoring and improving. Apart from its lovely interiors, Ragley is full of treasures. There are good paintings, including a landscape by Vernet, *The Raising of Lazarus* by Cornelis van Haarlem and many family portraits; the furniture is outstanding, and there is a fine collection of silver including a cruet set made by Paul Storr in 1804. The Prince Regent's bedroom contains a bed made in 1796. The mural in the south staircase hall, completed in 1983, was painted by Graham Rust.

☎ Alcester (0789) 762090/762455

11 m S of Redditch on A435

SP 0755 (OS 150)

Open mid Apr to early Oct daily exc M and F (but inc Bank Hol M) pm; gardens also am June to Aug

⊖ (1 m) P WC ⊟ (by appt at any time) D (gardens only) ♣ ♥ 🏠 ◆ ⚥ 🐕 🕴

Stoneleigh Abbey

Kenilworth, Warwickshire

Originally a Cistercian abbey and then an Elizabethan house, Stoneleigh was transformed during the early 18th century, and turned into an Italianate mansion by the 3rd Lord Leigh. The only part of the monastic building to survive is the 14th-century gatehouse; the rest was in ruins by 1561, and the site had to be cleared before Sir Thomas Leigh could begin to build the house where he and his descendants were to live. Until 1710 Stoneleigh remained an Elizabethan building, but three years earlier Edward Leigh had married an heiress, and now he began to plan a new house that would reflect the spirit of the age. His architect was Francis Smith of Warwick, who also planned and built the interior and designed the panelling of the ground-floor rooms. This now survives only in two rooms, the silk drawing room and the velvet drawing room, but the exterior remains virtually unchanged. The elaborate plaster work in the saloon was carried out in the 1780s, possibly to the designs of Cipriani, but no further alterations were made until 1836, when the long gallery was destroyed and replaced by the present ground-floor porch and entrance corridor. The house contains exceptionally fine furniture, much of it made for the house during Francis Smith's building, which took twelve years. There is also an excellent collection of paintings, mainly portraits, including one of Henry VIII and a small painting of Prince Charles Edward, the Young Pretender, who (legend claims) was smuggled out of the house in a beer cask in the guise of a mysterious 'Mr Fox'.

☎ Kenilworth (0926) 52116

4 m N of Royal Leamington Spa on A444, turn NW to National Agricultural Centre

SP 3171 (OS 140)

Open Easter and May Bank Hols (exc Good Fri); June to Aug daily exc T and S; Sept Su only: house 1300-1700, gardens 1100-1730

Ⓟ WC 🖼 🚻 ♿ 🛍 🍴 ◆ ☀ 🎎 ● 🎎 🧒

Aston Hall

Trinity Road, Aston, Birmingham, West Midlands

This fine Jacobean mansion was begun in 1618 for Sir Thomas Holte, 1st Baronet, and completed about 1635. The architect is not known, but it may have been John Thorpe, as there are two manuscript plans for it in his *The Book of Architecture*. It has been little altered, though it underwent some remodelling in the late 17th and early 18th centuries. It remained in the hands of the Holte family until 1817, when it was rented by James Watt, son of the famous engineer, who lived here to his death in 1848. He worked his family crest into the decorations in many places, and some of the upstairs rooms are decorated in the 19th-century taste – he was one of the first to commission furniture in the neo-Jacobean style. The house was empty for some years, and was then acquired by a private company, on whose behalf Queen Victoria opened it to the public in 1858 as a museum and place of public entertainment. It is now a branch of the Birmingham Museums and Art Gallery. The house contains some fine plaster ceilings, notably in the long gallery and the dining room, dating from about 1630, and several rooms have their original panelling. Although the Holte furniture and the other contents of the house were dispersed by sale in 1817 and 1849, it has now been furnished as far as possible in accordance with inventories of the 17th and 18th centuries, its contents being drawn from museums and private collections, and most of the rooms are open to the public. The house has recently been closed for restoration work, and was re-opened in April 1984.

☎ Birmingham (021) 327 0062

2½ m N of Birmingham city centre, W of A38

SP 0789 (OS 139)

Open end Mar to early Nov daily M-S 1000-1700, Su 1400-1700

♿ 🅿 WC 🔈 (limited access by appt) 🚻 ♦
🏛 ★ ◆ 🌿 ● (no flash) ♿

Hagley Hall

Hagley, near Stourbridge, West Midlands

The Palladian house and its park were created for Sir George, 1st Lord Lyttleton, in the 1750s. The Lyttletons had owned Hagley since 1564, but the old family home had by this time become rather dilapidated. The new house was designed by the 'gentleman' architect Sanderson Miller, who produced a plain building at the request of his patrons: Sir George wrote to him that 'we are pretty indifferent about the outside, it is enough if it is nothing offensive to the eye', and Lady Lyttleton, who had stronger views, vetoed anything 'Gothick'. The interiors, however, are superb, and show where their real preoccupations lay. The lovely plasterwork, the delight of the formal rooms, shows English Rococo at its height, and since a fire in 1925 all the decorations have been restored with great skill. Some furniture and paintings were destroyed, but many remain, including a pair of elaborate pier-glasses in the columned gallery, attributed to the famous carver Thomas Johnson. The finest of the rooms is undoubtedly the tapestry drawing room, which was unharmed by the fire. It was designed around a set of beautiful Soho arabesque tapestries woven in 1720, and the entire room is in the most delicate Rococo style, with carved mirror- and door-frames complementing the plasterwork, a painted ceiling, and French 17th-century chairs covered with tapestry. The park contains some fine garden buildings, the best being the Rotunda of 1747, Miller's 'Gothick' castle and the Temple of Theseus designed in the Greek style by 'Athenian' Stuart.

☎ Hagley (0562) 882408

7½ m NE of Kidderminster on A456, turn E at junction with A491

SO 9180 (OS 139)

Open Easter to end May Bank Hol Su and M, July to end Aug daily exc S 1400-1700

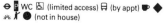 WC (limited access) (by appt)
(not in house)

Wightwick Manor

Wolverhampton, West Midlands

At first glance Wightwick (pronounced Witick) looks like one of the black-and-white timber-framed houses so common in Lancashire and Cheshire, but in fact it is late Victorian, and its main importance lies in its superb decorations by William Morris and other artists and craftsmen influenced by Ruskin and the Pre-Raphaelites. In 1887 Samuel Theodore Mander, a wealthy paint manufacturer, bought the old manor and employed the architect Edward Ould, a specialist in timber-framed buildings, to design a new house near the old one. The earlier buildings were given dressings of shiny red Ruabon brick, which was also used for the first stage of the new house. Six years later the more elaborate east wing was begun, with architectural details from Little Moreton Hall, spiral Tudor chimneys and intricate carving. The interior displays examples of all aspects of Morris' own designs – wallpaper, fabrics, carpets and tiles – as well as the work of other artists and craftsmen supplied by the firm – metalwork by W. A. S. Benson, tiles by William de Morgan and stained glass by C. E. Kempe. The house also contains paintings and drawings by Burne-Jones and other Pre-Raphaelites, and Morris' designs, as he intended, provide a setting for Jacobean furniture, Persian rugs and Chinese porcelain. The collection has been enlarged by the late Sir Geoffrey Mander and Lady Mander, who still lives in the house and is an expert on the period. The gardens, contemporary with the house and conceived as part of it, were laid out by Alfred Parsons.

☎ Wolverhampton (0902) 761108

3 m W of Wolverhampton on A454

SO 8698 (OS 139)

Open Mar to end Jan Th, S, Bank Hol Su and M 1430-1730; pre-booked parties at other times

⊖ Ⓟ WC ♿ (limited access)🚻 (by appt) ♠ ✳
✗ (exc Bank Hols) ● (with permision) NT

Buckinghamshire

Ascott, Wing, Buckinghamshire (tel Aylesbury [0296] 688242). 2 m SW of Leighton Buzzard on A418. SP 8922 (OS 165). Small 17th-century farmhouse extended in the 1870s and 1930s, when the house was in the possession of the Rothschild family. It contains Anthony de Rothschild's collection of paintings (notably 17th-century Dutch, and 18th- and 19th-century English), Chinese porcelain and French furniture. The 19th-century garden has many rare trees and attractive flower borders. Open late July to late Sept T-Su pm, also Bank Hol M (but closed following T) pm; garden only Apr to mid July and late Sept every Th and last Su in month pm. ⊖ 🅿 WC ♿ (limited access) ⊟ (by appt) ♣ ● (not in house) NT

Claydon House, Middle Claydon, Buckinghamshire (Steeple Claydon [029 673] 349). 3½ m SE of Buckingham on A413, turn SW. SP 7225 (OS 165). House built by the 2nd Earl Verney in 1754, of which only the west front survives. The interior has lavish Rococo woodcarvings by Luke Lightfoot, including a unique Chinese room and parquetry and wrought-iron staircase. A Florence Nightingale museum is found on the upper floor. Open S-W pm, Apr to end Oct. ⊖ 🅿 WC ♿ (limited access) ⊟ (by appt) ▼ ♣ 𝕏 (by appt) NT

Cliveden, Maidenhead, Buckinghamshire (tel Burnham [062 86] 5069). 1½ m E of Maidenhead. SU 9185 (OS 175). House in the style of an Italian Renaissance palace, built in the 1850s by Charles Barry for the Duke of Sunderland; the third house on the site since the Restoration. It has been owned by the Astor family, and the Duke of Westminster; it is now used as a college. The magnificent formal gardens reach down to the River Thames, with statuary and water gardens. House open Apr to end Oct Th and Su pm; gardens open all year daily am and pm. ⊖ 🅿 WC ♿ (by arrangement) ⊟ (by appt) D (on lead, grounds only) ♣ ▼ ◆ 𝕏 NT

Courthouse, Long Crendon, Buckinghamshire. 2 m N of Thame via B4011. SP 6909 (OS 165). 14th-century building, with two storeys and half-timbered, probably built as a wool-store, but used for manorial courts from the 15th century until recent times. Open S and Su and Bank Hols am and pm, W pm only, Apr to Sept. ⊟ ⚘

Dorton House, Dorton, Buckinghamshire (Brill [0844] 238237). 6 m N of Thame off A4011. SP 6713 (OS 165). Jacobean red-brick house built in 1626, now used as a school. The interior includes the original great hall with screens passage; other notable features include the ceilings, fireplaces and staircases. Open May to July and Sept pm; other times by appt. 🅿 WC ♿ (limited access) ⊟ (by appt) D ♣ ⊓ 𝕏

Manor House, Princes Risborough, Buckinghamshire. In town centre, opposite church off market square. SP 8003 (OS 165). Jacobean red-brick, hip-roofed house with the original oak staircase. Two rooms open for show in interior. Open W pm by appt. ⊖ 🅿 ♿ D (by arrangement) NT

Milton's Cottage, Chalfont St Giles, Buckinghamshire (tel Chalfont St Giles [024 07] 2313). 4 m SE of Amersham on A413, turn SW. SU 9893 (OS 176). Half-timbered 16th-century cottage, the home of John Milton in 1665-66, where he wrote *Paradise Lost* and part of *Paradise Regained*. The house contains Milton relics and a library of early books. Open T to S am and pm, Su pm, Feb-Oct, also Spring and Summer Bank Hols. ⊖ ♿ ⊟ ♣ ◆ ⚘ 𝕏

Nether Winchendon House, near Aylesbury, Buckinghamshire (tel Haddenham [0844] 290101). 7 m SW of Aylesbury, on A418, turn N. SP 7311 (OS 165). Medieval and Tudor manor house, reworked and given a Gothick façade in the late 18th century by Scrope Bernard. The house contains Tudor and later furniture and decorations, and a collection of objects relating to the Bernard family. Open May to Aug Th pm; also May and Aug Bank Hol weekends; parties at other times by appt. ▣▣❏ (by appt) ♣ ▼ (by appt) ✿ ⅄ ● (not in house) NT

Wotton House, near Aylesbury, Buckinghamshire. 6 m N of Thame, in the village of Wotton Underwood. SP 6816 (OS 165). House built in 1704 on the plan of Buckingham House (which was to become Buckingham Palace), with interior remodelled in 1820 by Sir John Soane, containing fine wrought-ironwork. The gardens were laid out by Capability Brown. Open W pm, Aug and Sept. ⊖▣❏ (by appt) ♣

Derbyshire

Ednaston Manor, Ednaston, Derbyshire (tel Ashbourne [0335] 60325). 6 m NE of Derby, off A52. SK 2341 (OS 128). House built by Sir Edwin Lutyens in Arts and Crafts style; the garden is of particular interest, with a large collection of shrubs and unusual plants. Open Easter to Sept daily (exc S) pm. ⊖▣WC▣❏ (by appt) ♣ ▼ (Su only or by appt) ★ ▩ ✿

Sudbury Hall, near Derby, Derbyshire (tel Sudbury [028 378] 305). At Sudbury, 6½ m SE of Uttoxeter on A50, turn S. SK 1532 (OS 128). Early 17th-century house built for the Lords Vernon, with exceptionally fine Restoration interiors, including woodcarvings by Grinling Gibbons, a richly decorated staircase by Edward Pierce and murals by Laguerre. There is also a museum of

childhood. Open Apr to Oct W-Su and Bank Hols pm. ⊖▣WC▣❏ D (not in house) ♣ ▼ ⼎ ◆ ⅄ ● NT

Leicestershire

Bede House, Lyddington, Leicestershire. 7 m S of Oakham on A6003, turn E. SP 8796 (OS 141). Mid-15th century palace for the Bishop of Lincoln, converted into a bedehouse or almshouse by Lord Burghley in 1602. It retains its 15th- and 16th-century decorations and painted glass in the bishop's apartment. Open Apr to Sept daily am and pm (Su pm only). ⊖▣

Donington le Heath Manor House, near Coalville, Leicester-shire (tel Coalville [0530] 31259). 13 m NW of Leicester, off A50. SK 4212 (OS 129). Small manor house of about 1280, now restored. It contains a kitchen with medieval equipment below the great hall. There is a medieval herb garden, and a medieval barn in the grounds. Open Easter to Oct W-Su pm, and Bank Hol M and T pm. ⊖▣WC▣♣▼ ★

Prestwold Hall, Loughborough, Leicestershire (tel Loughborough [0509] 880273). 2½ m SE of Loughborough. SK 5721 (OS 129). 19th-century Classical house set in extensive gardens, designed by William Burn for the Packe family. It contains interesting plasterwork and a collection of 18th-century English and European furniture. Open by appointment for parties (minimum of 20). ▣WC♣ EH

Wygston's House, Applegate St, Leicester, Leicestershire (tel Leicester [0533] 554100). In city centre, near St Nicholas' Circle. SK 5804 (OS 140). Medieval timber-framed house with an 18th-century brick façade, originally the home of the merchant Roger Wygston. It contains a museum of historical costumes, and a sequence of period interiors and a 1920s shop reconstructions. Open M-Th am and pm, Su pm only. Closed Christmas and Good Fri. ⊖ WC ⬚ ⊟ (by appt) ★ ◆ ⚹

Northamptonshire

Aynhoe Park, Aynho, Northamptonshire (tel Croughton [0869] 810659). 7 m SE of Banbury on A41. SP 5133 (OS 151). Jacobean mansion, partly rebuilt in the 1680s and again between 1707 and 1714. Sir John Soane made further alterations in the early 19th century, inside and out. The house has now been partially converted into flats. Open May to Sept W and Th pm. ⊖ 🅿 WC ⬚ D (at other times by appt) D (not in house) ♣ 👝 𝄃

Canons Ashby House, Canons Ashby, Northamptonshire (tel Blakesley [0327] 86004). 18 m SW of Northampton, on B4525. SP 5750 (OS 152). A small manor house, with restored garden including 18th-century terraces and walls. The house was originally part of an Augustinian priory, and was converted in the 16th century by the Dryden family. There are Elizabethan wall-paintings, and Jacobean plasterwork. Open Apr to end Oct W-Su and Bank Hols pm. ⊖ 🅿 WC ⬚ ⊟ (by appt) ♣ 👝 ⊓ ⚹ ● NT

Delapré Abbey, Northampton, Northamptonshire (tel Northampton [0604] 762129). 1 m S of town centre, off A508. SP 7559 (OS 152). A 12th-century Benedictine nunnery, converted into a private house in 1538. Most of the house dates from the 17th-18th centuries, and it is used as the Northamptonshire Record Office and Library. There is an attractive park. House open Th pm; park open daily am and pm. ⊖ 🅿 WC ⬚ ♣ ★

Hinwick House, near Wellingborough, Northamptonshire (tel Rushden [0933] 53624). 5 m SE of Wellingborough on A509 turn E. SP

9362 (OS 153). Attractive Queen Anne house, with notable 17th-century paintings, including works by Van Dyck, Lely and Kneller. There are displays of lace and tapestries, and of fashion from 1840 to 1940. Open Easter, Spring and Summer Bank Hols; other times by appt. 🅿 WC ⊟ (by appt) 👝 ⊓ ⚹ 𝄃 ● (not in house)

Holdenby House, Holdenby, Northamptonshire (tel Northampton [0604] 770786). 4 m NW of Northampton on A50 turn NW. SP 6967 (OS 152). The largest private house in England in Elizabeth I's reign, built by Christopher Hatton. The original garden and park can be seen, including fragrant and silver borders, and rare breeds of cattle and sheep. House open only by arrangement for parties; gardens open Apr to Sept Su and Bank Hol M pm (also July and Aug Th pm). 🅿 WC ⬚ ⊟ (by appt for house) D (on lead, grounds only) ♣ 👝 ◆ ⊛ 𝄃 (compulsory for house) ● (not in house) ⚸

Kelmarsh Hall, near Market Harborough, Northamptonshire (tel Maidwell [060 128] 276). 7 m S of Market Harborough on A508. SP 7379 (OS 141). Neo-Classical house built by James Gibbs around 1730, set in attractive grounds, and now restored to its original appearance. The interior contains exceptional stucco plasterwork, and fine 18th-century furniture. Open Easter to end Sept, Su and Bank Hols pm. ⊖ 🅿 WC ⊟ (by appt) D ♣ 👝 ⚹ 𝄃

Triangular Lodge, Rushton, Northamptonshire (tel Rushton [0536] 71076). 4 m NW of Kettering on A6003, turn E. SP 8383 (OS 141). Three-sided house built by Thomas Tresham in 1593-96, incorporating the symbolic number three as a consistent architectural motif (three sides, three storeys, three windows, etc). Open Apr to Sept M-S am and pm (Su pm only). 🅿 ⬚ (limited access) ⊟ ♣ ◆ ⚹ ●

Nottinghamshire

Carlton Hall, Carlton-on-Trent, Nottinghamshire. 7 m N of Newark on A1, turn E. SK 7964 (OS 121). Georgian house built by Carr of York in 1765, set in attractive grounds. There is a large and magnificent drawing-room, with elaborate plaster decoration. Open by appt, Apr to Oct. 🖩

Newstead Abbey, Linby, Nottinghamshire (tel Mansfield [0623] 793557). 9 m N of Nottingham on A60, turn E. SK 5353 (OS 120). 13th-century abbey, converted into a private house in the 1540s by the Byron family, and restored in the 1820s by John Shaw. The original cloisters, and the west front of the church, survive. The house contains a Jacobean saloon, a fine 19th-century hall, and rooms containing relics of the 6th Lord Byron (the poet) and other family treasures. There is an extensive and beautiful park. House open Easter to end Sept daily pm; gardens open all year daily am and pm. ⊖ (limited) 🖩 WC 🚻 (limited access) 🚻 D (grounds only) ♣ 🍽 🎋 ◆ ☀

Thoresby Hall, Ollerton, Nottinghamshire (tel Mansfield [0623] 823210). 9 m SE of Worksop on B6005, turn E. SK 6371 (OS 120). 19th-century house in the Elizabethan style, designed by Anthony Salvin. It replaced a Georgian house of the 1760s. The interior is decorated in Tudor fashion, but contains 18th-century furniture. The park is set in Sherwood Forest, and has woodland walks. Open Easter, Spring and Summer Bank Hol weekends and June to Aug S park am and pm, house pm only; parties by appt W and Th. 🖩 WC 🚻 🚻 (by appt W, Th) ♣ 🍽 ◆ ☀ 🏃 ● (none)

Thrumpton Hall, Nottinghamshire (tel Nottingham [0602] 830333). 9 m SW of Nottingham on A453, turn W. SK 5031 (OS 129). Jacobean house of

1607 built by Gervase Pigot, with a Restoration period staircase and carved and panelled saloon. There is a formal garden and extensive park. Open by appointment only on any day of the year, for parties of 20 or more. 🖩 WC 🚻 🚻 (only) D ♣ 🍽 🎋 ◆ ☀ 🏃

Wollaton Hall, Wollaton, Nottinghamshire (tel Nottingham [0602] 281333). 2½ m W of Nottingham on A609. SK 5339 (OS 129). Exceptional Elizabethan house, built in the 1580s by Robert Smythson for Sir Francis Willoughby . The façade is heavily ornamented and served as the model for Mentmore Towers in the 19th century. The house is now used as a museum of natural history, covering botany, geology, zoology and a herbarium, with special emphasis on the natural history of Nottinghamshire. The grounds contain a 19th-century iron-and-glass camellia house, and an industrial museum in the stable block. Open daily am and pm (Su pm only). ⊖ 🖩 WC 🚻 (limited access) 🚻 ♣ 🍽 🎋 ◆ 🏃 (by appt) ● (no flash) 🏃 🐕

Oxfordshire

Ardington House, near Wantage, Oxfordshire (tel Didcot [0235] 833244). 2½ m E of Wantage, off A417. SU 4388 (OS 174). Classical house of about 1720, built of grey stone and red brick. The hall is notable, and there is an Imperial staircase and panelled dining room with painted ceiling. The grounds contain fine cedar trees. Open May to Sept M and Bank Hols pm; parties at other times by appt. ⊖ 🖩 WC 🚻 🚻 (by appt) ♣ 🍽 🎋 🏃

Ashdown House, Ashbury, near Lambourn, Oxfordshire. 3½ m N of Lambourn, off B4000. SU 2882 (OS 174). Restoration period house built alone on the Downs, in the tall Dutch style, by William Winde. There is a carved four-storey staircase, hung with family portraits; the grounds are laid out to late 17th-century taste. Open Apr to end Oct W and S pm (exc Easter, Bank Hol M). Grounds open all year S-Th, dawn to dusk. ⊖ 🅿 🅿 (limited access) D (grounds only) 🚻 (by appt) ♣ NT

Buscot House, near Faringdon, Oxfordshire (tel Faringdon [0367] 20786). 3 m SE of Lechlade on A417. SU 2496 (OS 163). House of the 1780s, restored to its original condition in the 1930s. Its contents include paintings (including works by Rembrandt and Burne-Jones) and fine furniture. The grounds include an Italian water-garden designed in the early 20th century. Open Apr to Sept W-F (inc Easter F, S, Su) and 2nd & 4th S and Su pm each month. ⊖ (2¼ m) 🅿 WC 🚻 (by appt) ♣ 🗷 𝘟 ● (not in house) NT

Ditchley Park, Enstone, Oxfordshire (tel Enstone [060 872] 346). 9 m NW of Oxford on A34, turn W. SP 3921 (OS 164). The third great 18th-century mansion in Oxfordshire, designed by James Gibbs and with interior decoration by William Kent and Henry Flitcroft. It was the weekend headquarters of Sir Winston Churchill during the Second World War, and is now used as an Anglo-American Conference Centre. Open 20th to 31st July daily pm. 🅿 WC 🚻 ♣ ⚭ 𝘟 ● (not in house)

Greys Court, Henley-on-Thames, Oxfordshire (tel Rotherfield Greys [049 17] 529). 3 m W of Henley-on-Thames. SU 7283 (OS 175). Jacobean house set in the 14th-century walls of a medieval manor. There is a Tudor donkey wheel for raising well water, and the house contains some excellent Rococo plasterwork. The very extensive gardens include a recently opened maze. Open Apr to Sept; house M, W, F pm, garden M-S pm (exc Good Fri). ⊖ (2 m) 🅿 WC 🅿 (limited access) 🚻 (by appt) ♣ 🗷 NT

Kelmscott Manor, Kelmscot, near Lechlade, Oxfordshire. 4½ m W of Faringdon. SU 2163 (OS 163).Large Cotswold manor house that was the home of Arts and Crafts reformer and socialist William Morris from 1871 to 1896. Morris directed the restoration of the house. The contents include examples of his work. Open on written application (sae). 🅿 WC 🚻 (by appt) ♣ ⚭ ◆ ⚬

Nuneham Park, Nuneham Courtenay, near Oxford, Oxfordshire (tel Nuneham Courtenay [086 738] 551). 7 m SE of Oxford on A423, turn W. SU 5497 (OS 164). 18th-century Palladian villa in a landscaped park by the River Thames. There is a temple in the grounds. The house is now a conference centre. Open last two weekends in Aug pm, inc Bank Hol. ⊖ (1 mile) 🅿 WC 🅿 🚻 D ♣ 🗭 ⚭ ◆ ⚬

Stanton Harcourt Manor, Stanton Harcourt, Oxfordshire. 5 m SE of Witney on A415, turn E. SP 4105 (OS 164). Unusually well-preserved medieval buildings, including the 'Pope's tower', old kitchen and domestic chapel. There is a fine collection of paintings, furniture, silver and porcelain. Open Apr to Sept, alternate Th and S also Bank Hol M pm. 🅿 WC 🅿 🚻 D (on lead, grounds) ♣ 🗭 (not Th) ⚭ ◆ ⚬

Staffordshire

Ford Green Hall, Smallthorne, Stoke-on-Trent, Staffordshire (tel Stoke-on-Trent [0782] 534771). SJ 8850 (OS 118). 4 m N of city centre, off A53. Timber-framed yeoman's cottage of the 16th century, with furniture of the 16th to 18th centuries. Open M, W, Th, S am and pm, Su pm only. ⊖ 🅿 🅿 (by appt) ♣ ⚭ ★ ◆ ⚬ 𝘟 (compulsory)

Warwickshire

Anne Hathaway's Cottage, Hewlands Farm, Shottery, near Stratford-upon-Avon, Warwickshire (tel Stratford-upon-Avon [0789] 292100). 1 m W of Stratford. SP 1854 (OS 151). Half-timbered, thatched farmhouse, the home of Anne Hathaway before her marriage to William Shakespeare. There is an attractive cottage garden. One of the most popular sites in Britain. Open daily am and pm, exc Su am Nov to Mar. ⊖ P WC ♿ ⊟ ♣ ☕ (summer only) ⛩ ◆ ⁂ ✗ ● (not in house)

Baddesley Clinton, Solihull, Warwickshire (tel Lapworth [056 43] 3294). 9 m NW of Warwick on A41, turn W. SP 2071 (OS 139). Moated medieval manor house dating in part to the early 14th century, and little changed since Jacobean times. There are 120 acres of grounds. Open Apr to Sept W-Su and Bank Hols pm; Oct, S and Su pm only. P WC ♿ ⊟ (by appt) ♣ ☕ ◆ ⁂

Coughton Court, Alcester, Warwickshire (tel Alcester [0789] 762435). 7 m S of Redditch, on A435. SP 0860 (OS 150). 15th- and 16th-century house with an early 19th-century façade, the home of the Throckmorton family, which was one of the leading Catholic families in Britain. The house contains relics of the history of British Catholicism, including objects relating to Mary, Queen of Scots, the Gunpowder Plot and the Jacobites. The gatehouse dates from 1509. Open May to Sept daily exc M, F (but inc Bank Hols) pm; Apr and Oct S and Su pm; also Easter week to Th. ⊖ P WC ⊟ (by appt) ♣ ☕ ◆ ⁂ ✗ (by appt) ✦ NT

Farnborough Hall, near Banbury, Warwickshire. 8 m N of Banbury off A423. SP 4349 (OS 151). 17th- and 18th-century house, mostly rebuilt in the mid-18th century, with fine Rococo plasterwork. There is a terraced garden with wide views and garden temples. Open Apr to end Sept W, S and Bank Hol M, pm. ⊖ P WC ♿ ⊟ D (grounds only) ♣ ⁂ ✗ (by appt) ● (not in house) Terrace walk only Th, F, Su NT

Halls Croft, Old Town, Stratford-upon-Avon, Warwickshire (tel Stratford-upon-Avon [0789] 292107). In town centre. SP 2055 (OS 151). Fine Tudor house with walled garden, the home of William Shakespeare's daughter Susanna and her husband Dr John Hall. There is an exhibition of 17th-century medicine in the house. Open Apr to Oct daily am and pm (Su pm only); Nov to March M-S am and pm. ⊖ WC ♿ ⊟ ♣ ☕ ◆ ⁂ ●

Honington Hall, Shipston-on-Stour, Warwickshire (tel Shipston-on-Stour [0608] 61434). 15 m SE of Stratford-upon-Avon off A34. SP 2642 (OS 151). Elegant house built in 1682 for the merchant Henry Parker, remodelled inside in the 1740s by John Freeman in elaborate Rococo style, with fine plasterwork. Open late May to late Sept W and Bank Hol M pm. ⊖ P WC ⊟ (by appt) ♣ ✗ ● (not in house)

Mary Arden's House, Wilmcote, near Stratford-upon-Avon, Warwickshire (tel Stratford-upon-Avon [0789] 293455). 4 m NW of Stratford, off A34. SP 1658 (OS 151). early 16th-century timber-framed farmhouse, the home of Shakespeare's mother Mary Arden. It contains a range of 16th- and 17th-century furniture. There are out-buildings of the Tudor period, including a dovecote, and a collection of objects connected with old farming techniques. Open Apr to Oct daily am and pm (Su pm only); Nov to March M-S am and pm. ⊖ P WC ♿ ⊟ ♣ ☕ ⛩ ◆ ⁂ ✗ ● (not in house)

New Place and Nash's House, Chapel St, Stratford-upon-Avon, Warwickshire (tel Stratford-upon-Avon [0789] 292325). In town centre. SP 2055 (OS 151). Foundations of New Place, Shakespeare's last home, bought by him in 1597 as the second largest house in Stratford, and to which he retired in 1612. The house itself was destroyed in 1759, but an Elizabethan knot garden remains. Nash's House next door contains a display of furniture and local history. Open Apr to Oct daily am and pm (Su pm only); Nov to March M-S am and pm. ⊖ WC ⓦ ⊟ ♣ ◆ ⚹ ⚷ ● (not in house)

Packwood House, Hockley Heath, Warwickshire (tel Lapworth [056 43] 2024). 11 m NW of Stratford-upon-Avon on A34, turn NE to Lapworth. SP 1772 (OS 139). Timber-framed mid-Tudor house with 17th-century additions. The house was altered in the 19th century, and restored to its original form in the 1920s and 30s. It contains fine 16th- and 17th-century furniture and tapestries, and the grounds are well known for their topiary, with a formal Carolean garden and a yew garden representing the Sermon on the Mount. Open Apr to end Sept W-Su and Bank Hol M pm.; Oct S and Su pm only. ⓦ WC ⓦ (limited access) ⊟ (by appt) ♣ ⊓ ◆ ⚹ ⚷ (by appt) ● (no flash or tripods) NT

Shakespeare's Birthplace, Henley St, Stratford-upon-Avon, Warwickshire (tel Stratford-upon-Avon [0789] 204016). In town centre. SP 2055 (OS 151). Half-timbered house in which the poet was born in April 1564. The house contains an exhibition of rare Elizabethan items, and a BBC costume display. Open Apr to Oct daily am and pm; Nov to March daily am and pm (Su pm only). ⊖ WC ⓦ ⊟ ♣ ◆ ⚹ ⚷ ● (not in house)

Upton House, Edge Hill, Warwickshire (tel Edge Hill [029 587] 266). 11 m NW of Banbury off A422. SP 3645 (OS 151). Late 17th-century house, heavily reworked in the 1920s, containing a fine collection of *objets d'art* (including furniture, Brussels tapestries and porcelain) and paintings. There are attractive terraced gardens with several lakes.

210

Open Apr to end Sept M-Th pm, and some weekends (phone for dates). ⊖ ⓦ WC ⓦ ⊟ (by appt) D (on lead, grounds only) ⊟ (by appt) ♣ ◆ ⚸ ⚹ ⚷ ● (not in house)

West Midlands

Blakesley Hall, Yardley, Birmingham, West Midlands (tel 021-783 2193). 3 m E of city centre. SP 1285 (OS 139). Large, ornate timber-framed yeoman's cottage of 1575, with 18th-century outhouses. Several rooms are furnished as period settings, and have displays of local history and crafts. There is an attractive formal garden. Open daily pm closed Christmas, New Year. ⊖ WC ⓦ (limited access) ⊟ ♣ ⊓ ★ ◆ ⚹ ⚷ (by appt)

Moseley Old Hall, near Wolverhampton, West Midlands (tel Wolverhampton [0902] 782808). 5 m N of Wolverhampton on A449, turn E. SJ 9204 (OS 139). Small brick-clad timber-framed manor house of the early 17th century, refuge of the defeated future Charles II after the battle of Worcester (1651). The house contains 17th-century furniture, and a display about the Whitgrave family, the owners at that time. There is a formal 17th-century-style garden. Open mid Mar to mid Nov S, Su and Bank Hol M and T following pm; July to 6 Sept W, Th, F pm; open Su in Dec; parties at other times (not M, T) by appt. ⊖ ⓦ WC ⓦ (limited access) ⊟ (by appt) ♣ ⚑ ⚹ ⚷ ● (with permission) NT

Oak House, Oak Road, West Bromwich, West Midlands (tel 021-553 0759). ¼m W of town centre, off A41. SP 0092 (OS 139). Gabled 16th-century timber-framed house with a unique lantern-tower. The interior has much original panelling and wood carvings, and contains 16th- and 17th-century furniture. Open Apr to Sept daily am and pm (Th am only, Su pm only); Oct to March M-S am and pm (Th am only). ⊖ WC ⓦ (limited access) ⊟ ♣ ⊓ ★ ◆ ⚹ ⚷

Eastern Counties

Kingston
upon Hull

HUMBERSIDE

Grimsby

Old Hall,
Gainsborough

Lincoln

Doddington
Hall

Harrington
Hall

Gunby
Hall

Aubourn
Hall

LINCOLNSHIRE

Fulbeck Hall

Marston Hall

Fydell House

Holkham
Hall

Sheringham
Hall

Felbrigg Hall

Belton House

Sandringham
House

Grantham House

Houghton Hall

Blickling Hall

Woolsthorpe Manor

Medieval
Merchant's
House

Beeston
Hall

King's
Lynn

NORFOLK

LEICESTER-
SHIRE

Burghley
House

Peckover
House

Oakleigh
House

Norwich

Somerleyton
Hall

Peterborough

Hale's Hall

Elton Hall

NORTHAMPTON
SHIRE

CAMBRIDGESHIRE

Hinchingbrooke
House

Euston
Hall

Wingfield College

Island Hall

Bury St
Edmunds

Ixworth Abbey

Heveningham
Hall

Kimbolton
Castle

Cambridge

Anglesey

Ickworth

Angel Corner

Haughley Park

Glemham Hall

SUFFOLK

Otley Hall

Wimpole Hall

The Priory

Ipswich

BEDS

Kentwell Hall

Melford Hall

Christchurch
Mansion

Houghton House

Audley End

Gainsborough's
House

Belchamp
Hall

Castle
House

The Redoubt

aburn Abbey

HERTFORD-

Luton

Gosfield Hall

Colchester

Knebworth House

Paycocke's

Luton Hoo

Shaw's Corner

Layer Marney
Tower

Shalom Hall

Piccott's
End

SHIRE

ESSEX

Ashridge

Hatfield House

Harlow

BUCKS

Gorhambury

Chelmsford

Watford

Moor Park

GREATER
LONDON

Southend-on-Sea

BERKS

Gillingham

SURREY

KENT

Canterbury

Luton Hoo

Park St, Luton, Bedfordshire

Luton Hoo is a museum first and a country house second. The house has many virtues, but only a true architectural gem could rival the marvellous collection of works of art within. The house as it is today dates from 1903, although there had been a house at Luton Hoo for 700 years before this. In the 1760s Robert Adam was employed by the Earl of Bute to rebuild the existing house, and further alterations were made by Robert Smirke in about 1800. The large park was landscaped by Capability Brown. However, very little of the architecture of this period survives, as the house was gutted by fire in 1843. Sir Julius Wernher, South African diamond magnate, bought it in 1903, and employed the firm of Mewès and Davies, architects of London's Ritz Hotel, to remodel it and give it a new roof. They were responsible for its splendid Edwardian appearance: the fine white marble staircase gives an idea of the quality of their work, but their *pièce de résistance* is the dining room, with its polychrome marble and opulent gilded cornice framing the superb Beauvais tapestries. Half the house is now the family's private residence, while the rest houses the art collections. There are fine pieces of furniture, French tapestries and important paintings, including a late work by Altdorfer, but the real stars are the objects. The ivories, enamels, Renaissance bronzes and jewellery cannot be equalled by any other private collection. One of the most fascinating is a display of Imperial Russian objects, including several by Fabergé.

☎ Luton [0582] 22955

3 m SE of Luton on A6129 turn SW to Luton Hoo Park

TL 1018 (OS 166)

Open mid Apr to mid Oct daily exc M but inc Bank Hol M 1400–1745

● (not in house)

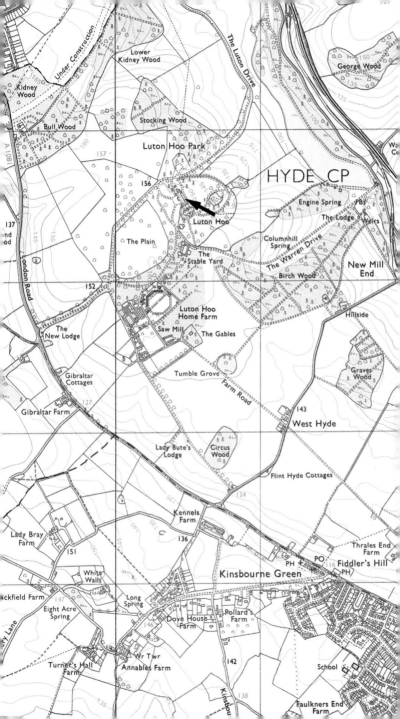

Woburn Abbey

Woburn, near Leighton Buzzard, Bedfordshire

The present Duke of Bedford, whose family has lived at Woburn since the 17th century, was the founder of the 'stately home business'; Woburn opened its gates to the throng in 1955, and has since become a thoroughly professional entertainment centre. The house, often ignored in favour of baboons and hippos, was built on the site of the cloister of the original Cistercian abbey, but it was not really lived in until Francis Russell took possession in 1619. Rebuilding began soon after, and the north wing became the private apartments. The west side, with its central pediment, was added by Henry Flitcroft between 1747 and 1761, and at the end of the century Henry Holland built the south and east sides. Until 1950 the house kept its original Cistercian quadrangular form, but sadly the entire eastern half, including Holland's east wing, had to be demolished because of dry rot. However, the house still retains an impressive and dignified appearance, and the lavish interior contains many treasures. The state rooms, designed by Flitcroft, include a Chinese room with a superb wallpaper brought from China in the 18th century, and most of the rooms have fine plaster ceilings. More of Flitcroft's work can be seen in his skilful remodelling of the long gallery. The house contains a superb collection of paintings, including works by Van Dyck, Velasquez and Reynolds as well as the famous Armada Portrait of Elizabeth I. The library is the finest of the Holland rooms, and Holland's pretty 'Chinese Dairy' can still be seen in the grounds.

☎ Woburn (052 525) 666

5 m N of Leighton Buzzard on A418, turn E at Woburn

SP 9632 (OS 165)

Open Apr to 1st Nov daily 1100-1700 (Su, Bank Hol M 1100-1830); Jan to end Mar S, Su 1100-1700

♿ (1½ m) 🅿 WC 🚻 D (guide dogs only in house)
🌲 🍴 🎡 ⛽ (not in house) 🚶

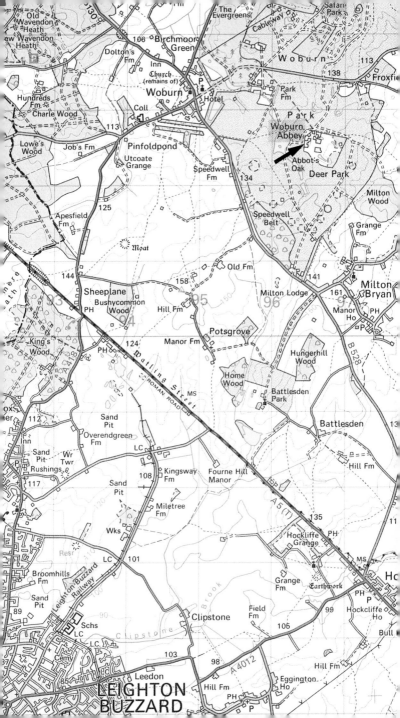

Anglesey Abbey

Lode, Cambridgeshire

The name of this house (formerly a priory of Augustinian canons, founded 1135) has nothing to do with the island of Anglesey, but comes from Angerhale, a fenland hamlet in the neighbourhood. The great attraction of the place lies not so much in the house – a comfortable residence built around the old priory chapter-house and day-room – as in the richness and variety of the art collection displayed within, and the generously laid out 20th-century park and gardens. A very rich owner, Huttleston Broughton, 1st Lord Fairhaven (1896-1966), was able to lavish a fortune on the collection and its setting. He built a library wing and a picture galleries block, adding a connecting bridge as late as 1955. To enjoy the furniture, *objets d'art*, paintings, engravings, tapestries and books and their extremely tasteful arrangement properly, visitors should allow themselves plenty of time and make use of the National Trust room-by-room guide. Lord Fairhaven's taste was catholic, ranging from the late Middle Ages to the 19th century, and cosmopolitan. A special interest of his was topographical views of Windsor Castle (the collection is now happy in the upper gallery). In the lower gallery are two paintings by Claude, and a selection from some twenty paintings by William Etty hang in the library corridor. The grounds are the other showpiece of Anglesey Abbey – an area of level fen entirely transformed with a breadth of conception that is almost unique in the 20th century.

☎ Cambridge (0223) 811200

4 m E of Cambridge on A1303 turn onto B1102 to Lode

TL 5262 (OS 154)

Open Easter S, Su, late Apr to late Oct W-Su and Bank Hol M 1330-1730

⊖ P WC & (limited access) ⊟ (W, Th and F, by appt) ♣ ☞ ⊼ ◆ ⚹ ● (by permission) NT

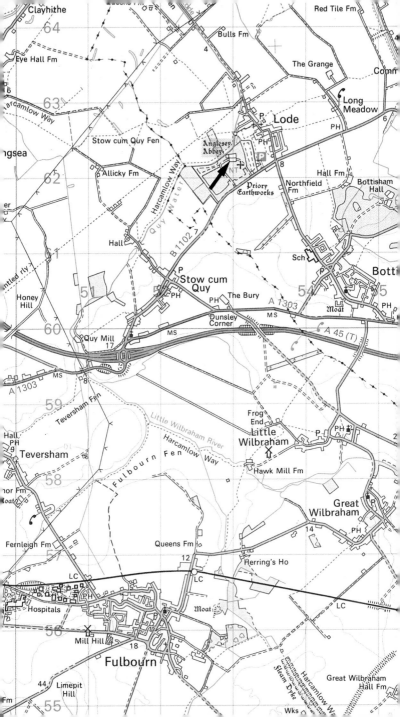

Burghley House

near Stamford, Cambridgeshire

No provincial building commemorates the power and prosperity of Elizabethan England better than Burghley House. Three storeys of mullioned and
transomed windows surround a courtyard and are surmounted by a single
spire and many cupolas and tall, coupled chimneys. The Renaissance
mansion clearly incorporated Italian, French and Flemish features, but
glimpsed now across the artificial lake or from the west across lawns the
weathered stones express English qualities: restraint, endurance and firmly
based aspirations. It serves as a fitting memorial to a great Englishman,
William Cecil, chief minister to Elizabeth I for forty years. Apart from all the
demands of high office, William Cecil seems to have been his own architect,
attending to every detail in the construction of the house, which was
completed in 1589. The interior was comprehensively reorganised between
1681 and 1700 by John Cecil, 5th Earl of Exeter, and from that time the Tudor
exterior contained some of the finest Baroque rooms in England. The state
apartments on the first floor are vast and numerous. The painted ceilings and
walls of many of the rooms are an outstanding feature. Most of these are by
Antonio Verrio; there is also work by one of his assistants, Louis Laguerre,
and by Stothard a hundred years later. The rooms are also filled with fine
furniture, pictures, tapestries, ceramics and carving.

☎ Stamford (0780) 52451

½ m S of Stamford on A43 turn onto B1081 and
then E

TF 0406 (OS 141)

Open Good Fri (pm) to early Oct daily 1100-1700,
exc S Horse Trials (Sept). Deer park open daily

⊖ (limited) P WC ♿ D (grounds only) ♠ ♣ ⊓
◆ ✶ ✗ ● (not in house)

Elton Hall

Elton, near Peterborough, Cambridgeshire

Elton Hall is a large house, built in a mixture of architectural styles which are best seen in the south, or garden front. This incorporates the oldest part of the building, the 15th-century tower and chapel built by the Sapcotes. This family lived here until 1600, and the tower bears their coat-of-arms. The Proby family, who own the house today, acquired it in the 17th century and Sir Thomas Proby built a relatively modest new dwelling, adding the north wing. The next major building campaign was in the period 1780-1815, when the gatehouse was joined to the house by a two-storey block, and the west front stuccoed and given a castellated Gothic look. Much of this work, however, was undone by the architect Henry Ashton who removed the Gothic parts of the west front, though he left them on the south front which still has a picturesque pseudo-medieval look. Finally, in the 1870s the 4th Earl of Carysfort built the central tower, a billiards room and new kitchen. The entrance hall has fine 17th-century panelling which may originally have come from the old Antwerp Town Hall; the marble hall and main staircase, designed by Ashton, are in the mid-Victorian 18th-century revival style; the chapel has 15th-century fan vaulting from the old chapel. The upper octagon room is a fine example of the Strawberry Hill 'Gothick' style, and the drawing room has an 18th-century ceiling. The library contains an exceptional collection of early bibles and prayer books, and there are some excellent paintings by Constable, Hobbema, Millais and Alma-Tadema.

☎ Elton (08324) 223

10 m SW of Peterborough on A605

TL 0892 (OS 142)

Open May to June: W; July to Aug: W and Su; May and Aug Bank Hol Su, M: 1400-1700

♿ 🅿 WC 🚻 (by appt) ♠ 🐕 ⴷ ◆ ⚘
✗ (compulsory exc Bank Hol weekends) ●

Island Hall

Godmanchester, Cambridgeshire

Looking at this stylish 18th-century mansion today it is hard to believe that in the late 1970s its owner could stand in the cellar and see daylight through the roof. The house was the home of the Baumgartner family (they later changed their name to Percy) from 1810 until 1943. Then it was requisitioned by the RAF, and later became the property of the local council and was converted into flatlets. It was allowed to become derelict, and then in 1977 a fire destroyed the south wing. The house seemed doomed, but amazingly many of the furnishings, including lovely carved fireplaces, survived intact, and when Simon Herrtage and his mother bought it in 1978 they were able to restore it to its 18th-century design. In 1983, their work complete, they put the house on the market, and it was reclaimed for the family by Christopher Vane Percy and his wife, who are continuing the work of restoration and refurnishing. The house had been built in 1750 by a gentleman with the unusual name of Original Jackson as a twenty-first birthday present for his son. The architect is not recorded, but the symmetry of the house suggests a professional hand, as do the well-balanced interiors, particularly the staircase hall with its soaring fluted columns. The muniments room, to the right of the entrance, has a collection of early photographs of the house and two family trees, both drawn up by ancestors of the owner, but one rather more truthful than the other.

1 m SE of Huntingdon on A604 at Godmanchester

TL 2470 (OS 153)

Open June to end Sept: Su, T and Th also May and Aug Bank Hol weekends 1430-1730

⊖ 🅱WC 🚻 ♠ ● ⚹ 𝒇 occasional concerts

Wimpole Hall

Arrington, near Royston, Cambridgeshire

This huge house, the largest in Cambridgeshire, set in an equally huge park, is notable for the astonishing number of architects and landscape gardeners it gave employment to during the course of its building. The original house, some of whose internal walls survive, was built between 1640 and 1670, the formal gardens were laid out at the end of the century, and the east and west wings were added by the architect James Gibbs for Edward Harley between 1713 and 1721. The library, also by Gibbs, was built in 1730 to house the famous Harleian collection. In 1740 Wimpole was sold to the 1st Earl of Hardwicke, for whom some further alterations were done; Sir John Soane made improvements to the interior for the 3rd Earl, and the 4th Earl employed the architect H. E. Kendall to build a new service wing and stables. The 5th Earl, nicknamed 'Champagne Charlie', sold the estate, and some time later it was bought by Captain Bambridge and his wife Elsie, daughter of Rudyard Kipling, who removed some of the less pleasing Victorian additions. The interiors are enormously rich and varied, Soane's yellow drawing room, and the south drawing room, which is a combination of Flitcroft and Gibbs, being two of the loveliest rooms. The great library, containing over 50,000 volumes, is most impressive, and the Baroque chapel, which fills the whole of the east wing, is entirely covered with painted decorations by Sir James Thornhill. The house contains excellent 18th-century furniture and paintings collected by Mrs Bambridge.

☎ Cambridge (0223) 207257
11 m SW of Cambridge on A603 turn NW to Old Wimpole
TL 3351 (OS 154)

Open early Apr to late Oct daily exc M, F 1300-1700, Bank Hol M 1100-1700

♿ (limited) 🅿 WC ♿ 🚻 (by appt) D (on lead, grounds only) ♣ 🍴 ⌂ ◆ ⚘ ● 🐕 🚶 NT

Audley End

Saffron Walden, Essex

Built round three sides of the medieval cloisters of the Benedictine Abbey of Walden, Audley End was the largest building solely of the Jacobean period ever built. It was constructed by Thomas Howard, Baron de Walden and later 1st Earl of Suffolk, between 1605 and 1614. Four wings rose to three storeys round an inner court; the scale can be judged from the long gallery, 240 foot in length, which occupied the first floor of the east wing. A huge outer court was built on the west side, a separate kitchen block to the north and stables beyond the river Cam. The house had cost a fortune, and debts led to Howard's disgrace. The house was bought by Charles II, but in 1701 was granted back to the Howards, who spent forty years reducing it to a more manageable size. The north and south wings of the outer court were demolished, as was the kitchen block. Further demolitions took place under Elizabeth, Countess of Portsmouth later in the 18th century. She also renovated the house and formed the nucleus of the collections seen there today. Towards the end of the 18th century her nephew employed Capability Brown to landscape the park and Robert Adam to reorganise and redecorate the interior. The west front is a fine example of the Jacobean style. The interior of the house is a mixture of Jacobean, 18th- and 19th-century styles, and contains fine collections of furniture, pictures and books.

☎ Saffron Walden (0799) 22399

1½ m W of Saffron Walden at Audley End

TL 5238 (OS 154)

Open Apr to 1st Su Oct daily exc M but inc Bank Hol M 1300-1700; in Aug open 1100 Su and Bank Hol M

⊖ (limited) 🅿 WC ♿ (limited access) 🚻 D (on lead) ♣ 🍴 🎋 ◆ 🐕 ⚹ 📷 (no flash) EH

224

Layer Marney Tower

Colchester, Essex

No one could forget the first sight of the enormous Tudor gatehouse at Layer Marney; it is quite simply astonishing. Although there are several such gatehouses in East Anglia, none other is on this scale. It was begun in 1520 by Henry Marney, who rose to wealth under Henry VIII, but he died in 1523 and his son two years later, so the huge, ambitious, Renaissance brick house of which the tower is the gatehouse was never completed. There is an east and a west wing, and an isolated south range, but no courtyard. The gatehouse itself is three storeys high, and the pairs of towers flanking the central part, with their seven tiers of windows, rise clear above the roofline and finish in terracotta crests. Visitors are admitted inside the tower, but not into the adjoining wing, where the present owners live. On the second floor there are some interesting documents relating to the house and its former occupants, and here the windows are of terracotta with winged cherubs' heads at the apex. Continuing up the west staircase, the visitor can ascend to the roof, where decorations on top of the tower can be viewed clearly. On one side of the garden is a long brick range which was originally stables, but has now been converted into a long gallery. This can be visited, as can the parish church, containing Marney tombs, on the other side of the garden.

☎ Colchester (0206) 330202

7 m SW of Colchester on B1022 turn S to Layer Marney

TL 9217 (OS 168)

Open Easter/Apr to end Sept Su and Th, also July and Aug T 1400-1800; Bank Hol M 1100-1800

🅿 WC ♿ (limited access) 🚻 (by appt) D (on lead, grounds only) ♣ ♟ ♦ ✗ 🏊 (by appt)

Paycocke's

West St, Great Coggeshall, Essex

The house, completed about 1505, is named after its original owner, Thomas Paycocke, one of the town's leading clothiers, whose family is represented by four tombstones in the church. It is an outstanding example of the type of half-timbered house built in some profusion in this part of the world by successful tradesmen and merchants. The house has been much restored, and the fabric between the timbers is now brick, but originally it would have been wattle and daub, woven slats of wood covered in mud or plaster. Paycocke's is famous for its lovely carved decoration, and the frieze on the horizontal beam at the base of the upper storey is particularly interesting, and contains the initials of Thomas Paycocke and his trade sign as well as two reclining figures, a head growing out of a flower and a baby diving into a lily. There are five oriel windows in the upper storey. The interior contains more carving, that on the ceiling joists of the hall being especially good. The overmantel, which incorporates the arms of the Buxton family, who succeeded the Paycockes at the house, was made up in the 1920s from old timbers and is a composite piece. The house is let to National Trust tenants, and some of the furniture is theirs, but the large oak pieces have been lent to the house. The attractive garden, which is full of interesting and unusual plants, is also open to the public, and provides a good view of the buildings at the back of the house, some of which are very old.

☎ Braintree (0376) 61305

6 m E of Braintree on A120

TL 8422 (OS 168)

Open Apr to mid Oct T, Th, Su and Bank Hol M 1400-1730; parties of 6 and over by appt

⊖ ⌖ 🚻 (by appt only) ● (not in house) NT

Ashridge

Berkhamsted, Hertfordshire

The history of Ashridge spans some 700 years, but the house owes its character to a major rebuilding between 1808 and 1814 by James Wyatt in the 'Gothick' style. After the Dissolution of the Monasteries, Ashridge passed to the Crown, and was one of the residences to which Henry VIII's children were frequently sent. Elizabeth I disposed of it in 1575, and in 1604 it was acquired by Thomas Egerton. He made many improvements to the old monastic building, and it was his descendant the 7th Earl of Bridgwater who carried out the 18th-century rebuilding. James Wyatt died before the work was completed, and it was finished by his nephew Sir Jeffry Wyattville. Another member of the Wyatt family, Matthew Digby Wyatt, remodelled all the main interiors in the Italian style between 1855 and 1863. In the late 19th century Lady Brownlow, a talented hostess, turned the house into a setting worthy of her glittering hospitality, and royalty, distinguished politicians, artists and men of letters were entertained here. Ashridge was sold, under the terms of the 3rd Earl Brownlow's will, in 1923, and a large part of the enormous park was bought by the National Trust, though the pleasure gardens, designed by Humphry Repton, still belong to the house. Only the library is still used for its original purpose, but the opulent decorations throughout are unchanged. The showpiece of the 19th-century work is the conference room with its marble fireplaces and pillars and painted ceiling, while the lovely chapel is an outstanding example of Wyatt's romantic style.

☎ Little Gaddesden (044284) 3491

5 m N of Berkhamsted on B4506, turn E to Ashridge Park

SP 9912 (OS 165)

Open few weekends during summer, phone for detailed information

🅿 WC ⊟ (by appt only) D (on lead, grounds only)
♠ ⚲

Gorhambury

near St Albans, Hertfordshire

This large Palladian villa built for the Viscount Grimston in 1777-84 by Robert Taylor has recently been refaced in Portland stone. Gorhambury, a former property of St Alban's Abbey, was bought by Sir Nicholas Bacon in 1561, and the ruins of the house he built here can still be seen in the park. It passed to his younger son, the famous Sir Francis Bacon, and the library of the present house contains many of his books. Bacon had no heirs, and the old house was bought by Sir Harbottle Grimston in 1652. The architect of the new house is best known for his relatively small villas, but at Gorhambury everything is very large; a massive stair leading to a vast portico and then to the great cube of the hall. Here two stained-glass windows of about 1620 showing plants and exotic scenes give an idea of the decoration of the old house. There is also an extremely fine carpet of 1570, and the walls are hung with portraits of 17th-century figures. In the dining room there are portraits of Sir Francis Bacon and Sir Harbottle Grimston, and the ballroom contains several lovely portraits and still-lifes by Sir Nathaniel Bacon, nephew of Sir Francis. In the yellow drawing room, where everything is 18th century, there is a portrait by Reynolds, and a fine chimneypiece designed by Piranesi and bought in Rome. The library contains three mid-16th-century heads in painted terracotta of Sir Nicholas Bacon, his wife and son. Also in the library are photocopies of some of the earliest printed editions of Shakespeare's plays, which were found here and are now in the Bodleian Library, Oxford.

☎ St Albans (0727) 54051

In W outskirts of St Albans on the A414 turn NE to Gorhambury

TL 1107 (OS 166)

Open May to Sept Th 1400-1700; reduced rates for parties by appt

🅿 🚻 (by appt) ⚹ 𝄚 (compulsory)

Hatfield House

Hatfield, Hertfordshire

The old Hatfield Palace, built by the Bishop of Ely in about 1497, was retained by Henry VIII after the Dissolution of the Monasteries as a residence for his daughters. Most of it was demolished in 1608 by Robert Cecil, chief minister under James I, who acquired it in an exchange of houses with the King. He chose a more prominent site on which to build his grand new mansion. His is a thoroughly Jacobean building, confident and lavish. Stone quoins and parapets, large mullioned and transomed widows serve to break up the expanse of brickwork. A soaring white clock tower rises in three stages over the central block and the massed chimneys of the two wings, and gold leaf originally adorned the domes and turrets. Cecil built the house in the expectation of royal visits, and it contains separate state apartments for the King and Queen, connected by the long gallery that runs the whole length of the south front. On the ground floor was a great hall rising two storeys, and below the King's state rooms an apartment for Cecil's own use. He did not live long enough to enjoy the house, but his descendants have lived there ever since. The house remains today much as it was originally. The great hall has a marble floor, wood and plaster ceiling, Brussels tapestries and screened wooden gallery, and contains two important portraits of Elizabeth I, whose association with Hatfield permeates the house. The grand staircase leading to the King's rooms in the east wing is a masterpiece of Jacobean carving, and an early example of a cantilevered staircase.

☎ Hatfield (30) 62823

21 m N of London on A1, opposite Hatfield BR station

TQ 2308 (OS 166)

Open late Mar to mid Oct daily exc M, Good Fri: T-Su 1200-1700, Su 1400-1730, Bank Hol M 1100-1700

⊖ P WC ⬧ 🚻 D (grounds only) ♣ 🐾 ◆ 🐾
🚍 ⳩ (compulsory, exc Su, Bank Hols) ●

Knebworth House

Knebworth, Hertfordshire

The theatrically romantic exterior of Knebworth House, with its turrets and pinnacles and its gargoyles silhouetted against the sky, was the creation of the Victorian novelist Edward Bulwer-Lytton, 1st Lord Lytton. But the 19th-century decoration conceals a house dating back to Tudor times. Sir Robert Lytton began building at the end of the 15th century, and successive generations added, subtracted, altered and redecorated. The visitor enters on the west side, the only remaining Tudor wing, into an entrance hall given its present form in the early part of this century by Sir Edwin Lutyens. The banqueting hall, the great hall of the Tudor building, ranks as one of the most beautiful rooms in England. It is now entirely 17th century, the oak decoration of the ceiling hiding the Tudor open-timber roof. Sir Rowland Lytton's great oak screen is a perfect example of the native Jacobean style, while the other walls show Classical influences from Italy. The library was Bulwer-Lytton's and his novels still fill the shelves. The mid-19th-century Jacobean-style staircase leads to his study. The state drawing room, originally the presence chamber leading to the long gallery (now demolished) is a superb example of Victorian High Gothic decoration, the work of John Crace. The Gothic furniture is similar to that which Crace made for Pugin; walls and ceiling are painted with heraldry, and the window contains a stained-glass portrait of Henry VII.

☎ Stevenage (0438) 812661
3½ m S of Stevenage on B197 turn W at Knebworth for Old Knebworth
TL 2320 (OS 166)

Open Apr and May S, Su, Bank Hol M and school hols; late May to mid Sept daily exc M 1200-1700

⊖ Ⓟ WC ♿ (limited access) D (grounds only)
🛏 (by appt) ♠ ➡ ◆ ※ ✗ ● ⚘

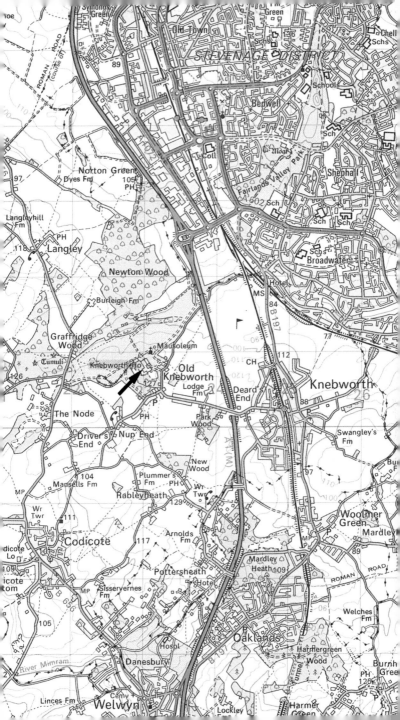

Moor Park

Moor Park Golf Club, Rickmansworth, Hertfordshire

The splendid 18th-century mansion of Moor Park is now a golf club, but the house is still open to the public. In the 1680s a brick house was built for the Duke of Monmouth, which was much admired by contemporaries, but in the 1720s it was completely restyled by Benjamin Styles. The painter Sir James Thornhill apparently had a hand in the design until he quarrelled with Styles, but the principal architect was Giacomo Leoni, who was responsible for the magnificent Corinthian portico on the west front. The huge and perfectly proportioned entrance hall has a painted and gilded ceiling with a painted dome in imitation of that in St Peter's, Rome, and is the work of Thornhill, while the wall paintings of mythological subjects were done by the Italian artist Amiconi. The elaborate plaster decoration is probably also of Italian workmanship, and the staircase leading to the gallery is decorated with more mythological paintings by Sleker. The lounge at the back of the house is known as the Thornhill room because its superb painted ceiling was believed to have been done by Thornhill, but some experts now think it was painted by Verrio, in which case it must have formed part of the decorations of the older house, since Verrio returned to Italy in 1707. The room also has two carved fireplaces and fine panelling, both pre-Georgian, and the dining room has a coffered ceiling with painted decoration by Cipriani, part of a programme of 'beautification' carried out by Sir Lawrence Dundas, who bought the property in 1763.

☎ Rickmansworth (0923) 776611
2 m SE of Rickmansworth on A404 turn NE to Moor Park
TQ 0793 (OS 176)

Open throughout year M-F 1000-1600, S 1000-1200; visitors must report to reception

⊖ ▮ 🅿 🖃 (by appt) ♣ ★ ⚬ 🎋 ●

Belton House

near Grantham, Lincolnshire

This house of 1684-88 arouses little gasps of pleasure from architectural writers, and really does deserve its label as a 'perfect house from the age of Wren'. Relatively little altered from the time when it was built at the turn of the reigns of Charles II and James II, its sombre panelled and carved interiors and formal garden layout preserve the feeling of that period. The Brownlow family have lived on this site since about 1640, and the present Lord Brownlow lives in part of the house still. Some changes were made by James Wyatt and Sir Jeffry Wyattville for successive owners, but the overall impression of a home of a wealthy gentleman of late Stuart England remains intact. The paintings – many of them hung on panels framed by carved swags of high craftsmanship in the style of Grinling Gibbons – are in keeping with the original period. A family chapel of no mean size is part of the house, and it contains a richly decorated reredos which would not look out of place in St Paul's Cathedral. The main floor of the chapel is fitted with contemporary pews for the servants and grooms, while an upper gallery, garlanded with carved fruit and flowers, accommodated the owner's family on armchairs.

☎ Grantham (0476) 66116

3½ m N of Grantham on A607 turn E to Belton

SK 9239 (OS 130)

Open Apr to end Oct W-Su, Bank Hol M 1300-1730, grounds 1100-1830; parties at other times by appt

⊖ P ⬚ (limited access) ⬚ (by appt) D (on lead, grounds only) ♣ ⬛ ◆ 🎍 🌿 🕆 ● ⚲ NT

Harrington Hall

Spilsby, Lincolnshire

A charming 17th-century house, long and well proportioned, built in a lovely mellow pinkish brick. In the 14th century the manor belonged to the Coppeldykes, who rebuilt the medieval house in 1535; the Elizabethan porch of the present house is a survival from this period. In 1673 the Tudor house was bought by Vincent Amcott, who entirely rebuilt the main part of it, leaving only the porch tower with its curious pilasters dating from about 1660. He also inserted most of the panelling. Despite its length it is, rather surprisingly, only one room thick, and only the ground-floor rooms are shown to visitors. The hall is part of the Tudor building, but was remodelled in the 1720s, the low, wide, elliptical arch taking the place of the former screen. The fine Doric panelling was also brought in at this time, and there is a William and Mary clock made by John Blundell in a lovely inlaid case. The oak staircase, carved with hops and wheat-ears, is also of the 1720s, but was rebuilt in 1951 owing to dry rot; fortunately the carving was not harmed. Most of the rooms are panelled, with the drawing-room panelling, put in before 1700, being the oldest. This room also contains the best furniture – some fine mainly 18th-century pieces – though there is good furniture in the other rooms too. The panelling in the dining room is 18th century, and has recently been painted Indian red. The delightful semi-formal walled garden is believed to be the original of Maud's garden in Tennyson's poem. The poet lived nearby and was in love with Rosa Baring of Harrington Hall.

☎ Spilsby (0790) 52281

7 m E of Horncastle on A158 turn NE after Hagworthingham to Harrington

TF 3671 (OS 122)

Open Easter to Sept Th 1400-1700; gardens open Easter to July some Su, Bank Hol M 1400-1800

P WC ♿ (by appt) 🚻 (by appt) D (not in house) ♣ ☕ (limited opening) ◆ ❀ ⚘ 🅟 ●

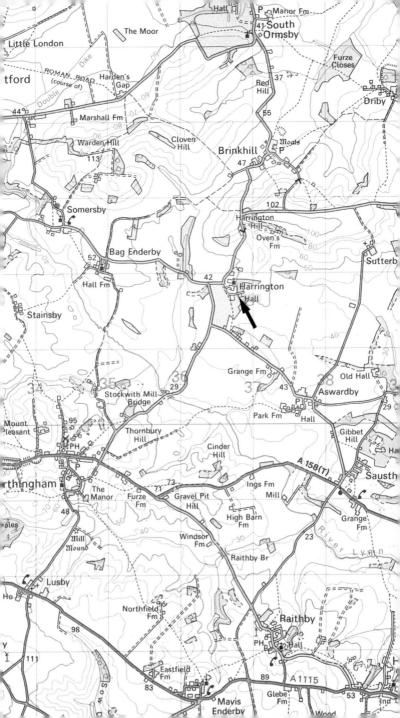

Blickling Hall

Blickling, Norwich, Norfolk

The Cecils' Hatfield House spawned Blickling: Robert Lyminge built both, and both buildings were status symbols of new men of the Jacobean age – one the Lord Treasurer, Robert Cecil, 1st Earl of Salisbury, the other the Lord Chief Justice, Sir Henry Hobart. Blickling was built between 1616 and 1625, and the entrance front, with its flanking, gabled outbuildings, is still as it was when finished in that year – a perfectly preserved forecourt of the early Stuart period. The other side of the house was rebuilt in the 1770s but as a scheme that harmonised with the features of the earlier parts (unusual for that time of sweeping 'modernisation' in the Palladian mood), and by using the same red brick established continuity on all four fronts of the building. Changes were made at the same time indoors, but much of Robert Lyminge's work survives – in particular the 120-foot-long gallery with its original plasterwork ceiling of allegories and armorials. The main staircase was moved to a different position and enlarged into a double-flight, but the original material was re-used for much of it, and some extra figures were carved to make up the full number needed – one of the intriguing features of Lyminge's staircase having been these soldiers and other characters standing on the newel-posts. The later work was done for the 2nd Earl of Buckinghamshire, who had brought back from serving as ambassador at St Petersburg a large tapestry of Tsar Peter the Great at the Battle of Poltava, for which hanging space on a wall of appropriate size was to be provided.

☎ Aylsham (0263) 733084

15 m N of Norwich on A140 continue onto B1354 for 3 m

TG 1728 (OS 134)

Open mid Apr to late Oct T, W, F, S, Su, Bank Hol M 1300-1700; closed Good Fri

⚑ WC ♿ ☐ (by appt) D (on lead, grounds only) ♣ ☕ ⛱ ◆ ⚘ NT

Holkham Hall

Wells, Norfolk

An outstanding Neo-Classical mansion on the west Norfolk coast, the house, designed by William Kent for Thomas Coke, was begun in 1734. It consisted of a central block and four wings, and remains almost unaltered, both inside and out. The exterior was always restrained, almost severe, but Coke filled the interior with books, sculptures and other works of art collected during his travels in Europe. Inside the 19th-century porch and unassuming north door, the marble hall is rightly famous. It is Coke's and Kent's idea of a Roman temple set down in the English countryside. The magnificent ceiling rises the full height of the building, and marble stairs lead to a peristyle of fluted Ionic columns of Derbyshire alabaster. The saloon has another magnificent, complex ceiling, and the walls are hung with their original Genoa velvet. The furniture, as elsewhere in the house, was designed for the room by Kent. In some half a dozen rooms the pictures are mainly Thomas Coke's collection, and the statue gallery was built specifically for the busts and statues he brought back from Italy. His famous collection of books is still contained in the Long Library. Another famous Coke, Thomas William, 'Coke of Norfolk', one of the greatest of agricultural pioneers, carried out his work at Holkham. The scale of the estate must have been gratifying to him: the park extends to 3000 acres, and is surrounded by a stone wall nine miles long.

☎ Fakenham (0328) 710227
2 m W of Wells-next-the-Sea on A149 turn S of Holkham
TF 8842 (OS 132)

Open June to end Sept Su, M, Th, also W in July and Aug 1330-1700; Bank Hol M 1130-1700

⊖ (limited, 1 mile) P WC 🔄 (limited access) 🚻
D (on lead, grounds only) ♦ 🍴 ⊼ ◆ 🎨 ✄ ●

Houghton Hall

Houghton, King's Lynn, Norfolk

Houghton Hall, one of the most splendid Palladian houses in Britain, was built by Sir Robert Walpole on the site of a Jacobean house, and completed in 1735. The original designs were prepared by Colen Campbell in 1721 but revised by Thomas Ripley, who added the four domes, and chose the attractive and hard-wearing Aislaby sandstone, which was brought by sea from Whitby. For the interior Walpole engaged a third architect, William Kent, which proved a masterstroke as he was responsible for everything – the marble fireplaces, the carved woodwork, most of the murals and a large part of the furniture. The state rooms are on the first floor, up the great mahogany staircase lined with mural paintings by Kent. The centrepiece of the state rooms is the 'stone hall', a perfect 40-foot cube in which all the decoration is on the grandest possible scale. The lovely carved ceiling frieze is by Atari; the reliefs over the fireplace and door are by Rysbrack; the chairs were made for the house (the green velvet upholstery is original); there is a 16th-century Persian carpet and an Aubusson of the Louis XV period. The other state rooms are equally splendid: the cabinet room has 18th-century Chinese hand-painted wallpaper; the tapestry drawing room was designed around a set of Mortlake tapestries depicting the Stuart kings and queens; the green velvet bedchamber has a state bed designed by Kent. The most sumptuous room of all is the saloon, with its luxurious furniture, walls covered in crimson Genoa silk velvet, and its carved and gilded ceiling.

☎ East Rudham (048522) 569

16 m NE of King's Lynn on A148 turn N at Harpley to New Houghton

TF 7928 (OS 132)

Open Easter Su to end Sept Su, Th, Bank Hols 1300-1730; reduced rates for parties by appt

⊖ (1 m walk) 🅿 WC ♿ 🍴 (by appt) D ♣ ☕
🍴 ◆ ♒ ✗ (by appt) ● ☂

238

Oakleigh House

Swaffham, Norfolk

Oakleigh House, standing in the north corner of Swaffham market place, has no pretensions to grandeur; it was built as a comfortable farmhouse, and remains so today. It was built on the site of a dwelling occupied by John Chapman, the 'Swaffham Pedlar', who is reputed to have found buried treasure following a prophetic dream. The house was probably begun in the late 16th century and continued in the 17th, and the imposing façade dates from the mid-18th century. The estate was evidently still in the hands of the Chapman family in 1658, as in 1893 a carved oak door-head was discovered with the inscription 'J.C. 1658 E.C.' (John and Edward Chapman), but since that time it has passed through many hands, becoming the home of the Grammar School headmaster in 1949 and a sixth-form centre in 1975. In 1982, after remaining empty for two years, it was auctioned and is now once again a home. The house is interesting as, being a humbler dwelling than the grand country houses, it gives an idea of ordinary middle-class life. The stone-flagged floor in the attractive reception hall is original, and there is an attractive dado of carved pine of the 18th century. The fine Jacobean staircase, built in 1620, is similar to that at Blickling Hall, though simpler, and there are 17th-century doors and door jambs at the top of the house. The massive fireplace in the living room, three times filled in, and plastered over before restoration, is now a working fireplace again.

☎ Swaffham (0760) 24280

In Swaffham on Market Place

TF 8209 (OS 144)

Open Apr to end Sept Th and F 1400-1700; parties by appt at other times

⊖ ▯ ◰ ♣ ◗ ⚹ 𝙆 (parties only, by appt)
● (permission required)

Sandringham House

Sandringham, Norfolk

The Sandringham estate was bought by the Royal Family in 1861 when the future Edward VII came of age, but by the time of his marriage to Princess Alexandra the old house had already become inadequate, and today's sprawling scarlet-brick house was built in 1870. The architect was A. J. Humbert, who had built Prince Albert's Mausoleum at Windsor, and the style is solidly neo-Jacobean. Visitors are shown five main rooms, and the front door opens directly into the saloon, built on the pattern of a Jacobean great hall, with a minstrels' gallery at one end and the walls hung with 17th-century tapestries. The main drawing room, in the French style, has plaster and carved panelling, and is all in white with a painted ceiling. The paintings on the walls, both here and in the saloon, are mainly Victorian, with many portraits, and the atmosphere of the rooms is Edwardian, although they are used by the Royal Family today. The tapestries in the dining room, two of which were woven from Goya cartoons, were given by the King of Spain in 1876, and there are many other treasures. The sporting pictures and trophies on show in the lobby and ballroom corridor reflect the particular feature of life at Sandringham – Edward VII developed the estate into one of the finest game reserves in the country. In his day the beaters wore special uniforms, which are sometimes shown in the old stable block. This also contains gifts presented to Her Majesty the Queen, big game trophies, vintage royal cars and a vintage fire engine. The Victorian gardens are still very attractive.

☎ King's Lynn (0553) 772675

8½ m NE of King's Lynn on A149 turn E

TF 6928 (OS 132)

Open Apr to late Sept M-Th 1100-1645, Su 1200-1645. Closed late July to early Aug

⊖ 🅿 WC 🔄 🚻 ☕ 🍴 ⛤ ◆ 🐕 ❋
● (not in house) ⛏ ⛏

Christchurch Mansion

Soane St, Ipswich, Suffolk

This delightful red-brick Elizabethan house, now a museum, was built on the site of an Augustinian priory in 1548 for Edmund Withipoll, a successful merchant with cultural leanings. The original plan was the usual Tudor E-shape, and the diamond pattern of burnt blue bricks is a feature of several East Anglian houses. Additions were made to the west wing before 1600, and a fire in the 17th century necessitated further improvements. The dormers with Dutch gables in the attic storey date from this period, and the hall was repanelled in the later Stuart style. Christchurch has been lived in by three successive families. Edmund Withipoll's granddaughter married Colonel Leicester Devereux, and the house passed into his family in 1645. Nearly a century later it was sold to Claude Fonnereau, whose family lived here until 1894. Demolition was threatened, but Felix Cobbold, a banker, bought it and presented it to Ipswich Corporation. It has been a museum since 1896, entirely due to his generosity and foresight. From 1929 new galleries were built attached to the main building, including the Wolsey Art Gallery, where the best of the Ipswich Museum's art collections are shown. The Wingfield Room, with panelling from the town house of the Wingfield family, and the two Tudor rooms, have been entirely reconstructed from local timber-framed houses. The other rooms are furnished in a variety of period styles, and contain a fascinating and diverse collection of furniture and fittings from Tudor times to the 19th century.

☎ Ipswich (0473) 53246

In Christchurch Park, in centre of Ipswich

TM 1645 (OS 169)

Open daily throughout year exc some Bank Hols 1000-1700 (dusk in winter)

⊖ 🅿 ♣ ★ ◆ ⅍ 🅰 🏌 (foreign languages by appt) ● (permit required) 🚹 🚺

Gainsborough's House

Sudbury, Suffolk

In 1723 John Gainsborough, the father of the painter Thomas Gainsborough (1727-88), bought two small, old-fashioned houses and united them, adding an elegant Georgian façade. The house remained in the Gainsborough family for many years, and only underwent one major alteration, in the 1790s, when the whole of the back of the house was remodelled in the 'Gothick' style. It was bought by the Gainsborough House Society in 1958, after having been a hotel for some considerable time, and it was opened to the public as a museum in 1961, restored as far as possible to its appearance of 150 years ago. By 1971 it had become established with a good permanent loan collection of Gainsborough's paintings, 18th-century furniture and some of the artist's personal possessions, but the rooms are also interesting in themselves, as they show clearly the different periods of architecture and decoration. The entrance room, part of the 15th-century house, has some of its original timbering, though the fireplace and alcove date from about 1600. The parlour is 16th century with fittings of this date, and here there are drawings and watercolours by Constable as well as three of Gainsborough's Bath portraits. The bedroom, like the entrance room, is part of the Tudor house, and a panel of wattle-and-daub has been exposed to show the construction. The room contains a fine 18th-century tallboy and portraits of three of the artist's contemporaries. The exhibition rooms, at the back of the house are used for temporary exhibitions of work by modern artists.

☎ Sudbury (0787) 72958

In Gainsborough St, in centre of Sudbury

TL 8741 (OS 155)

Open Easter to Sept T-S 1000-1700, Su 1400-1700. Oct to Easter closes 1600 exc Gd Fri, 24 Dec-1 Jan

⊖ P WC ☒ (ground floor only) ⊟ (by appt) ♣ ◆
🕱 ● (not in house)

Heveningham Hall

Stowmarket, Suffolk

Sir Gerald Vanneck inherited the estate in 1777 and immediately decided to demolish the existing small 18th-century house and build a grand mansion. His architect for this extremely fine Palladian house, one of the best in the country, was Sir Robert Taylor, who had started his career as a sculptor. The house consists of a central block with pillars, rising from an arcaded basement and with wings on each side, also pillared. The park, with its fine lake, was laid out by Capability Brown in 1780. Heveningham's greatest glory is its interior, by James Wyatt. Taylor had begun his designs for the interior before the building was finished, so he cannot have been pleased to hear that a much younger man was to be entrusted with the whole of the interior (Wyatt was then thirty-four and Taylor sixty-six). But whatever ill-feeling there may have been, the rooms, all in Wyatt's own version of the Adam style, lighter and airier than Adam's, are a triumph, and he cleverly transformed the empty, box-like rooms into gentle, curving spaces by means of apses, semi-domes, niches and coves. The hall at Heveningham is one of the most beautiful rooms in England, pale blue, with a screen of golden-yellow *scagliola* (imitation marble) columns at either end, a vaulted ceiling with white plaster roundels and a patterned marble floor. The other rooms are scarcely less fine, and the saloon has painted decoration by Biagio Rebecca. The interior is perfectly preserved, and the entrance hall, Etruscan room and library still have their original furnishings.

☎ Ubbeston (098 683) 355

Telephone for details of opening

5 m SW of Halesworth on B117

TM 3573 (OS 156)

Ickworth

Horringer, Bury St Edmunds, Suffolk

The extraordinary building, modelled on an earlier circular house, Belle Isle on Lake Windermere, was the creation of the enormously rich traveller and art collector Frederick Hervey, Bishop of Derry and 4th Earl of Bristol. Sadly he did not live to see his masterpiece; begun in 1795, it was not completed until 1830. The Earl-Bishop died in Italy in 1803 when the house was only half built. The original design was by an Italian architect, Mario Asprucci, but the work was carried out by the Sandys brothers, Francis and the Rev. Joseph. Ickworth is a huge house, with the great central domed rotunda linked to the two wings by long, curving passages intended as art galleries. When the Earl-Bishop died, his son, who became the 1st Marquess of Bristol, wanted to demolish it as it was so impractical, but in the end he completed it, reversing the original scheme so that the east wing became the living quarters and the great rotunda rooms were used for art displays and receptions. The last major alterations were made in the 1900s, when Reginald Blomfield was employed to remodel the east wing and make improvements to the rotunda, and the interior of the ground floor, with three vast reception rooms grouped round a top-lit staircase-well, owes its appearance to his alterations. The rooms contain good 18th-century furniture acquired by the 1st Marquess and a few of the Earl-Bishop's own treasures, including John Flaxman's sculpture *Fury of Athemas* in the hall. The Pompeian Room, decorated by J. D. Crace in 1879, contains the original architect's model of the house.

☎ Horringer (028 488) 270

2½ m SW of Bury St Edmunds on A143 at Horringer

TL 8161 (OS 155)

Open May to end Sept daily exc M and Th; Apr and Oct, S and Su, also Bank Hol M 1330-1730

🅿 WC ♿ 🚻 (by appt) ♠ ♣ ⌂ ◆ ⊞ ⚒ ⚘
⚘ ● NT

Somerleyton Hall

near Lowestoft, Suffolk

This splendid and imposing Victorian palace was the creation of Sir Morgan Peto, a clever ex-bricklayer turned railway contractor. He was only thirty-three in 1844 when he bought the existing 17th-century house and proceeded to bring it up to date. The original house was meant to provide the inspiration for the rebuilding, and Peto employed the sculptor John Thomas to turn it into a 'Jacobean mansion'. What they produced, however, is the purest red-brick Victorian, albeit with hints of Italian and French styles. There is a very Italian tower on one side, and the French look comes partly through the use of the soft, pale Caen stone, which Thomas favoured because it was easy to carve. There is a great deal of carved ornament on the building, and the stone connecting screen between the wings in the French Renaissance style, is most elaborately carved. The main rooms with their oak panelling seem rather sombre in contrast to the bright exterior. In the oak parlour the 17th-century panelling survives, and the staircase hall and dining room have some of the original features. But the Victorian rooms are the most impressive, particularly the entrance hall, which has dark oak woodwork relieved by marble panels, Minton floor tiles, a painted stained-glass dome and stuffed polar bears. Sadly, Sir Morgan's business failed and he had to sell the house in 1863. It was bought by the Crossleys, later Lords of Somerleyton. They recently redecorated and refurnished the dining room, for which a carpet was woven by John Crossley and Sons, the family firm.

☎ Lowestoft (0502) 730224

8 m NW of Lowestoft on B1074

TM 4997 (OS 134)

Open Easter Su to end Sept Th, Su and Bank Hol M, also Tu and W in July and Aug 1400-1730

⊖ (1½ m walk) P WC ⟨⟩ 🖫 ♣ 🍴 ㅠ ◆
🌿 🎋 (by appt) ● (not in house) ⚘

Bedfordshire

Houghton House, Ampthill, Bedfordshire. 8 m S of Bedford, to E of B530. TL 0339 (OS 153). Ruined red-brick mansion built for the Dowager Countess of Pembroke 1615-21, with heavily decorated stone centrepieces on the north and west fronts. Inigo Jones has been attributed as the architect, and John Bunyan is said to have been inspired by it. The house was dismantled in the 1790s; only the outside walls remain. Open at any reasonable time. ⊖ ⅃ 冏 D ★ EH

Cambridgeshire

Hinchingbrooke House, Huntingdon, Cambridgeshire (tel Huntingdon [0480] 51121). ½ m W of Huntingdon on A604. TL 2271 (OS 153). 13th-century Benedictine nunnery dissolved in 1538; the house was acquired by the Cromwell family and was extended in the 16th century, then passed to the Montagu family in 1627. It was again extended in the 1660s, and restored in the 1820s. The house is now a comprehensive school. Open Apr to Aug, Su 1400-1700; also Bank Hol M. ⊖ 圓 WC ⅃ 冏 (by appt) ♦ ⬤ ⁂ ✗

Kimbolton Castle, Kimbolton, Cambridgeshire (tel Huntingdon [0480] 860505). 9 m NW of St Neots on A45. TL 1067. Tudor manor house, the home of Katherine of Aragon in the 1530s, and mostly remodelled in the early 18th century by Vanbrugh, with murals by Pellegrini and a gatehouse by Robert Adam. It is now used as a school. Open Easter, Spring and Summer Bank Hol Su and M pm; also late July to end Aug Su pm. ⊖ 圓 WC 冏 ♦ ⁂ ✗ (by appt)

Peckover House, North Brink, Wisbech, Cambridgeshire (Wisbech [0945] 583463). Close to town centre, on river bank. TF 4509 (OS 143). 246

Town house built in 1722, with a fine interior, displaying rich Rococo carvings, plasterwork and many other details. There is a notable Victorian garden, containing many rare and exotic trees (including fruit-bearing orange trees), and outstanding 18th-century stables. Open May to Sept, S to W 1400-1730; 11-30 Apr, 1-18 Oct S, Su and Bank Hol M only (1400-1730). ⊖ WC ⅃ (garden only) 冏 (by appt) ♦ ⬤ ⁂ ⬤ (not in house) NT

Essex

Belchamp Hall, Belchamp Walter, Sudbury, Essex (tel Sudbury [0787] 72744). 5 m SW of Sudbury. TL 8240 (OS 155). Queen Anne brick-built house with furniture and family portraits of the 17th and 18th centuries. Open by appt only, May to Sept T, Th and Easter, Spring and Summer Bank Hol M 1430-1800. 圓 WC ⅃ 冏 (by appt) ♦ ⬤ (by appt) ✗ (by appt)

Castle House, Dedham, Essex (tel Colchester [0206] 322127). 7 m NE of Colchester on A137, turn N. TM 0632 (OS 168). Georgian house, from 1919 to 1959 the home of Alfred Munnings, former President of the Royal Academy. There is a large collection of his paintings and sketches on many subjects, notably of horses. Open 1st Su in May to 1st Su in Oct W, Su and Bank Hol M 1400-1700; also Aug Th and S pm. ⊖ 圓 WC ⅃ (limited access) 冏 (by appt) D (with permission) ♦ ⬤ (not in house)

Gosfield Hall, Halstead, Essex (tel Halstead [0787] 472914). 4 m NE of Braintree on A1017, turn N. TL 7729 (OS 167). Unusual house built in the mid-16th century and with 18th-century additions and façades. The Tudor long gallery has notable panelling. Open May to Sept, W and Th pm. ⊖ 圓 WC ⅃ (limited access) 冏 (by appt) ♦ ⁂ ✗

Shalom Hall, Layer Breton, near Colchester, Essex. 6 m SW of Colchester on B1022, turn S. TL 9418 (OS 168). Victorian house which contains a collection of 17th- and 18th-century French furniture and porcelain, and several notable 18th-century portraits. Open Aug M-F am and pm. ▣ ★

Hertfordshire

68 Piccott's End, near Hemel Hempstead, Hertfordshire (tel Hemel Hempstead [0442] 56729). 2 m N of Hemel Hempstead on A4146, turn E. TL 0509 (OS 166). Medieval cottage once used as a hostel for pilgrims, and with a remarkable 15th-century religious wall-painting. The house contains interesting early kitchens, and in 1826 became the first cottage hospital in Britain. It now houses a collection of early medical instruments. Open early March to late Nov daily 1000-1800. ⊖ (limited) ▣ ⯊ (limited access) ⊟ (by appt) ◆ ꝭ ● (not in house)

Shaw's Corner, Ayot St Lawrence, Hertfordshire (tel Stevenage [0438] 820307). 1 m NE of Wheathampstead. TL 1916 (OS 166). Early 20th-century small house, the home of George Bernard Shaw from 1906 to his death in 1950. It has been preserved as it was during his lifetime, and contains a large collection of his belongings; his plays are produced in the grounds. Open Apr to end Oct M-Th 1400-1800, Su 1200-1800. ▣ WC ⯊ (limited access) ⊟ (by appt throughout year) D (on lead, car park only) ◆ ⽊ NT

Lincolnshire

Auburn Hall, near Lincoln, Lincolnshire (tel Bassingham [052 285] 270). 9 m S of Lincoln on A46, turn SE. SK 9262 (OS 121). 16th-century house attributed to the Smythson family, with a notable Jacobean carved staircase and

panelled rooms. Open July and Aug W 1400-1800, Su 28 June & 12 July (1987) 1400-1800; other times by appt. ⊖ ▣ ⊟ (by appt) ◆ ꝭ

Doddington Hall, Doddington, Lincolnshire (tel Lincoln [0522] 694308). 3 m S of Lincoln on A1180, turn NW onto B1190. SK 8970 (OS 121). Large Elizabethan house built in about 1600 by Robert Smythson, with a gabled Tudor gatehouse. The interior has been redecorated to a Georgian taste, and contains a rich variety of furniture and porcelain. There is a large park and walled rose garden. Open Easter M, May to Sept W, Su and Bank Hols 1400-1800. ▣ WC ⯊ (ground floor) ⊟ (by appt) ◆ ⯌ ◆ ⽊ ꝭ (by appt) ⯊

Fulbeck Hall, near Grantham, Lincolnshire (tel Loveden [0400] 72205). 14 m S of Lincoln on A607. SK 9450 (OS 121). A mainly 18th-century house in a large garden, the home of the Fane family for 350 years. The interiors contain interesting collections of furniture and paintings. Open June to Aug F and S, also Easter M and May Bank Hol M 1400-1700. ⊖ ▣ WC ⯊ ⊟ (by appt) D (on lead, grounds only) ◆ ⯊

Fydell House, South St, Boston, Lincolnshire (tel Boston [0205] 51520). In town centre. TF 3343 (OS 131). House built in 1726 for William Fydell, three times mayor of Boston. It is now used as a college. It has notable plasterwork, panelling and a fine carved staircase. Open term time M-F 1000-1600, other times by appt (phone [0205] 68588). ⊖ ▣ ⊟ (by appt) ★ ꝭ (by appt) ● (permission required)

Grantham House, Castlegate, Grantham, Lincolnshire. In town centre. SK 9136 (OS 130). House dating from the 14th century and much altered over the centuries. The grounds run down to the River Witham. Open Apr to Sept W and Th 1400-1800. ⊖ ▣ ⯊ ⽊ ● (not in house, no flash) NT

Gunby Hall, Burgh-le-Marsh, Lincolnshire. 10 m NW of Skegness on A158. TF 4666 (OS 122). House built in about 1700 in the style of Sir Christopher Wren, the home of the Massingberd family. There is some fine 17th-century furniture, and portraits by Joshua Reynolds. The garden is formal and walled. Open Apr to Sept Th 1400-1800; other times by appt. ⊖ 🅿 WC 🅰 (garden only) 🚻 ♦ ⚹ ✗ (by appt) ● (not in house) NT

Marston Hall, Grantham, Lincolnshire (tel Loveden [0400] 50225). 6 m NW of Grantham on A1, turn NE. SK 8943 (OS 130). 16th-century manor house modified in the 1720s and built by the Thorolds family. There is an ancient garden and an 18th-century Gothick gazebo. Open on occasional Sundays, summer only; other times by appt. 🅿 WC 🅰 (limited access) 🚻 D (on lead) ♦ ⚹

Old Hall, Gainsborough, Lincolnshire (tel Gainsborough [0427] 2669). In town centre. SK 8190 (OS 112). Excellent 15th- and 16th-century brick and timber manor house with medieval kitchens and associated rooms (perhaps the best survivng in England), and a huge great hall with original oriel window and roof. The Pilgrim Fathers used to meet here and John Wesley preached several times at Old Hall. There is a small museum of archaeological finds from the locality. Open throughout year M-S 1000-1700, Su 1400-1700 (closed Su Nov to Easter). ⊖ 🅿 WC 🚻 (by appt) ⚭ (T pm only) ♦ ⚹ ✗ (by appt)

Woolsthorpe Manor, near Grantham, Lincolnshire (tel Grantham [0476] 860338). 11 m S of Grantham on A1, turn W to B6403 for ¼ m then W. SK 9224 (OS 130). Small farmhouse of about 1620, the birthplace of Sir Isaac Newton in 1642. It was here that he returned during the plague years of 1665-66

and according to legend conceived the Theory of Gravitation in the orchard. Open Apr to Oct W-Su and Bank Hol M 1300-1700. ⊖ 🅿 WC 🅰 (limited access) 🚻 (by appt) D (on lead, grounds only) ♦ ⚹ ✗ ● (no tripods or video) NT

Norfolk

Beeston Hall, Beeston St Lawrence, near Wroxham, Norfolk (tel Horning [0692] 630771). 14 m NE of Norwich on A1151. TG 3321 (OS 134). 18th-century house built by William Wilkins, with fantasy Gothick exterior and a mixture of Gothick and Classical Georgian interiors. It is set in a picturesque landscaped park. Open Easter to mid Sept F, Su, Bank Hol M 1400-1730, also W in Aug. ⊖ 🅿 WC 🅰 (limited access) 🚻 (by appt) ♦ ⚭ ✗ (parties only) ● (not in house)

Felbrigg Hall, near Cromer, Norfolk (tel West Runton [026 375] 444). 2 m SW of Cromer on A148, turn S. TG 1939 (OS 133). Jacobean house built in about 1620, with the west front added in the William-and-Mary style. The interiors are mainly 18th and 19th century, and there is a fine library, with books from Dr Johnson's collection. There is a walled garden, orangery and landscaped park. Open mid Apr to 1st Nov daily (exc T, F) 1330-1730. 🅿 WC 🅰 🚻 (by appt) ♦ ⚭ ⊓ ◆ ⚹ ● (not in house) ⚘

Hale's Hall, Loddon, Norfolk (tel Raveningham [050 846] 395). 12 m SE of Norwich on A146. TM 3797 (OS 134). Fortified medieval manor house, with a great hall of the 1470s restored in the 1970s. The ruins and moat have ben excavated, and there are demonstrations of East Anglian crafts. Open Bank Hol Su, M May to Aug only, 1400-1730; other times by appt. 🅿 WC 🅰 (ground floor) 🚻 (by appt) ♦ ⚭ ⊓ ⚿ ⚹ ✗ (by appt) ● (not in house)

Medieval Merchant's House, King St, King's Lynn, Norfolk (tel King's Lynn [0553] 772454). In town centre. TF 6120 (OS 132). Medieval house built in the 14th century, now with a Georgian façade. The interiors are of the 17th, 18th and 19th centuries. Open June to Aug F and Su, also Bank Hol M, and July to Aug T, am and pm. ⊖ ⬓ (limited access) ⊟ (by appt) D (on lead) ♣ ☛ (by appt for parties) ⚲ ⚑ (compulsory)

Sheringham Hall, Sheringham, Norfolk (tel Sheringham [0263] 733471). ¼ m W of town centre on A149, turn S. TG 1342 (OS 133). Early 19th-century house and garden designed by Sir Humphry Repton and his son. The grounds include a long rhododendron drive. House open by appt only; grounds open May to June M-S 1000-1800; mid-May to mid-June Su 1400-1800. ⊖ ⬓ (limited access) ⊟ (by appt) D (on lead, grounds only) ♣ NT

Suffolk

Angel Corner, Angel Hill, Bury St Edmunds, Suffolk (tel Bury St Edmunds [0284] 63233 ext 227). In town centre. TL 8564 (OS 155). Queen Anne house, containing a large collection of clocks and watches of many styles and dates. Open daily 1000-1700, Su 1400-1700; closed Christmas, Easter. ⊖ ⬓ ⊟ (by appt) ☛ ★ ⚲ ⚑ (by appt) ● (permission required for interiors of clocks)

Euston Hall, Euston, Thetford, Suffolk. 4 m SE of Thetford on A1088. TL 8978 (OS 144). House built in the Restoration period and altered in the 1750s. Much of the house was destroyed in the 20th century. What remains contains a fine collection of 17th-century portraits. There is a landscaped park, the work of William Kent and John Evelyn. Open June to end Sept Th pm. ⬓ WC ⬓ (limited access) ⊟ ♣ ☛ ⊞ ◆ ⚲ ⚑ (by appt) ●

Glemham Hall, Woodbridge, Suffolk (tel Wickham Market [0728] 746 219). 13 m NE of Woodbridge on A12. TM 3459 (OS 156). Elizabethan house remodelled in the early 18th century in a severe manner. There is a staircase of oak inlaid with walnut, and much 18th-century panelling. There is some notable japanned furniture, and pieces by Hepplewhite and Sheraton. The walled garden is attractive. Open Easter to end Sept Su, W and Bank Hol M 1430-1730. ⊖ ⬓ WC ⬓ (limited access) ⊟ (by appt) ♣ ☛ ◆ ⚲ ● (not in house)

Haughley Park, near Wetherden, Suffolk (tel Elmswell [0359] 40205). 4 m NW of Stowmarket on A45, turn NW. TM 0062 (OS 155). Jacobean manor house, built on the E-plan and gabled. The garden front dates from 1800. The interior has been totally and accurately reconstructed in the 1960s. The gardens include woods and rhododendron walks. Open May to Sept T 1500-1600. ⊖ ⬓ WC ⬓ ⊟ (by appt) D (on lead) ⚑ (by appt for parties)

Ixworth Abbey, near Bury St Edmunds, Suffolk (tel Pakenham [0359] 30374). 8 m NE of Bury St Edmunds, on A143. TL 9370 (OS 155). 12th-century Augustinian priory, with some of the 13th-century cloisters surviving. There is a timber-framed prior's lodging of the late 15th century, and a private house mainly dating from the 17th century, was built on the site of the abbey. Open throughout year by appt only. ⊖ ⬓ ⊟ ☛ ⚑ (by appt)

Kentwell Hall, Long Melford, Suffolk (tel Sudbury [0787] 310207). 5 m N of Sudbury on A134, turn NW. TL 8647 (OS 155). Red-brick Tudor house on the E-plan with a moat, and approached by an avenue of limes planted in the 1670s. The interior was reworked in the 1820s, and is being refurbished in the 1980s. The restoration works may be viewed as part of the tour of the house. There is a 15th-century moat house and the grounds contain unusual trees. Open July to Sept W-Su 1400-1800; Apr to June W, Th and Su 1400-1800; also Bank Hol weekends 1200-1800, but last week June and 1st two weeks July S, Su: 1100-1700, M-F school parties only. ⊖ ▣ WC ⬥ (limited access) ⊟ (by appt) ♣ ☞ ◆ ⚶ ⟋ (by appt)

Melford Hall, Long Melford, Suffolk (tel Sudbury [0787] 7000). 3 m N of Sudbury on A134. TL 8646 (OS 155). Turreted brick house begun in 1554, built for Sir William Cordell. There is an 18th-century drawing room, and the interior was partly reworked in the 1810s, with a fine library. The contents include Chinese porcelain, 18th- and 19th-century naval paintings, and a collection of objects relating to Beatrix Potter, who was a regular visitor to the house. There is a Tudor pavilion in the garden. Open early Apr to end Sept W, Th, Su, and Bank Hol M (also June to Aug S) 1400-1800. ⊖ ▣ WC ⬥ ⊟ (by appt) ♣ ⚶ ● (not in house) NT

Otley Hall, Otley, near Ipswich, Suffolk. 9 m N of Woodbridge on B1079, turn NE. TM 2056 (OS 156). Moated 15th-century hall, the home of the Gosnold family for 250 years. It has fine timbers, herring-bone brickwork, panelling and frescos. Open Easter, Spring and Summer Bank Hol Su and M. ⊖ ▣ ⊟ (by appt) ☞

Priory, The, Water St, Lavenham, Suffolk (tel Lavenham [0787] 247417). In town centre, 6 m NE of Sudbury on B1115. TL 9149 (OS 155). A complex timber-framed building of the 13th to 16th century, with Jacobean staircase and fireplaces. The building has recently been restored, and contains a collection of photographs showing the progress of this work. Paintings, drawings and stained glass by Ervin Bossanyi. Open Easter to end Oct M-S 1030-1230, 1400-1730 (closed Su). ⊖ ▣ WC ⬥ (limited access) ⊟ (by appt) ♣ ☞ ◆ ⚶ ⟋ (by appt) ● (no flash)

Wingfield College, Eye, Suffolk (tel Stradbroke [037 984] 505). 7 m SE of Diss. TM 2277 (OS 156). 14th-century hall, the home of the Wingfield family for over 700 years. It contains a medieval cloister, gallery and lodgings range, with Elizabethan and Georgian interiors and an 18th-century façade. Open Easter to end Sept S, Su and Bank Hols 1400-1800. ▣ WC ⬥ (limited access) ⊟ (by appt throughout year) ♣ ☞ ⚶ ⟋ (by appt)

The North

Dundee

TAYSIDE

FIFE

Edinburgh

LOTHIAN

BORDERS

○ Berwick-upon-Tweed

DUMFRIES
&
GALLOWAY

Preston Tower 🏛

Howick Hall 🏛

🏛 Cragside House

NORTHUMBERLAND

Wallington House 🏛 🏛 Meldon Park

Belsay Hall 🏛

Castletown House

○ Carlisle

George Stephenson's Birthplace

🏛 Seaton Delaval Hall
Newcastle-upon-Tyne

TYNE & WEAR

🏛 Washington Old Hall

1	Belle Isle
2	Hill Top
3	Shandy Hall
4	Newburgh Priory
5	Sutton Park
6	Manor House, Ilkley
7	Harewood House
8	East Riddlesden Hall
9	Brontë Parsonage
10	Temple Newsam
11	Bolling Hall
12	Oakwell Hall
13	Shibden Hall
14	Red House
15	Platt Hall
16	Fletcher Moss
17	Wythenshawe Hall
18	Dunham Massey

Hutton-in-the-Forest 🏛

🏛 Wordsworth House

CUMBRIA

🏛 Dalemain

DURHAM

CLEVELAND

Middlesbrough
Ormesby Hall

Dove Cottage 🏛 🏛 Rydal Mount
Townend
🏛 1
Brantwood 🏛 2   🏛 Abbot Hall
Rusland Hall 🏛   🏛 Levens Hall

Kiplin Hall 🏛

Bedale Hall 🏛

Swarthmoor Hall   Holker 🏛 Hall

Osgodby Hall 🏛 🏛 3   🏛 Nunnington Hall
Norton Conyers 🏛 🏛 4   Castle Howard
NORTH YORKSHIRE
Markenfield Hall 🏛   🏛 Newby Hall 🏛 5

Ebberston Hall 🏛

Sewerby Hall 🏛

Leighton Hall

○ Lancaster

Broughton Hall 🏛

LANCS

Beningbrough Hall 🏛
Ripley Castle   York ○   🏛 Treasurer's House
Stockeld Park
🏛 Bramham Park

Sledmere House   Burton Agnes Hall

Norman Manor House

Browsholme Hall 🏛

Whalley Abbey 🏛

🏛 8
🏛 9   🏛 11 Leeds 🏛 10

Lotherton Hall

HUMBERSIDE

Lairgate Hall 🏛

Burton Constable 🏛
Blaydes House 🏛

Chingle Hall 🏛
Samlesbury Hall 🏛
Hoghton Tower 🏛

Gawthorpe Hall

WEST YORKS
13 🏛 🏛 12
14

Carlton Towers

Maister House 🏛
Wilberforce House 🏛 🏛

Kingston upon Hull

Meols Hall 🏛
Rufford Old Hall 🏛

Astley Hall 🏛 Turton Tower
Smithills 🏛 🏛 Hall i'th'Wood
Hall   Heaton Hall 🏛

🏛 Nostell Priory

Cannon Hall 🏛   Cusworth Hall 🏛   Old Rectory

🏛 Normanby Hall

MERSEYSIDE

Manchester 🏛 Foxdenton House
Ordsall Hall 🏛   Newton Hall
18 🏛 🏛 15 16
17

SOUTH YORKSHIRE

Croxteth Hall 🏛
Bluecoats Chamber 🏛
Speke Hall 🏛   Liverpool

Sheffield

CHESHIRE

DERBYSHIRE

NOTTS

LINCOLNSHIRE

CLWYD

POWYS

Stoke-on-Trent

STAFFORDSHIRE

Nottingham

SHROPSHIRE

LEICESTERSHIRE

Belle Isle

Windermere, Cumbria

This appropriately named house, standing on the largest island in Lake Windermere, is very unusual, being the first and almost the only house to be built on a circular plan. It was built in 1774, and copied at Ickworth some twenty years later. Belle Isle was designed by a then unknown architect, John Plaw, for a Mr English, who had just bought the island. Built of a very hard stone from the nearby Ecclerigg quarry, it is three storeys high and a true circle, 54 feet in diameter, with only the portico projecting. Plaw wanted the house to look perfect from every angle, and coped with the problem of deliveries and waste disposal by planting the house in a small sunken area, which also concealed the kitchens and servants' accommodation. In 1800 a new kitchen was built at the back, slightly marring the perfection of the plan, but this extension blends in quite well. Visitors are shown round the ground floor rooms only, which are small, with fine Adam plasterwork. An elegant staircase rises in the middle of the house. When it was first built, the house did not seem to accord with contemporary taste and aroused a storm of protest – the poet Wordsworth condemned the destruction of the natural beauty of the island – and Mr English sold the entire property for £1,720 to Miss Isabella Curwen in 1781 (he had spent £6,000). Much of the furniture, made in the 18th century by the well-known firm of Gillow, comes from the Curwens' other house, Workington. Bills for the furniture making are on show in the house.

☎ Windermere (09662) 3353
On Lake Windermere 2 m S of Windermere, boat from Bowness Promenade nr Cockshott Point
SD 3996 (OS 97)

Open May to Sept Su-Th 1030-1700; parties at other times by appt

♿ WC 🚻 (by appt) D (on lead, grounds only)
🌹 🍴 ⊓ ◆ ⚲ 🍼 (parties by appt) ● 🏃

252

Holker Hall

Cark-in-Cartmel, Grange-over-Sands, Cumbria

Formerly the home of the Dukes of Devonshire, Holker Hall dates originally from the 17th century, but it was altered and refaced in 1840 by Lord Burlington to give it a Gothic look, with tall ornamental chimneys, gables and mullioned windows. This part, known as the old wing, is closed to the public. In 1871 a fire broke out, destroying the entire west wing before it could be brought under control, and soon afterwards the 7th Duke began plans for rebuilding. A grand new wing was designed by Paley and Austin of Lancaster, who have been described by Pevsner as 'the best architects living in the country'. The new wing, built of red sandstone in the Elizabethan style, is their outstanding work; the craftsmanship is superb and there are many delightful details. The interior of the house, which contains a mixture of Georgian and later furniture, is spacious, and visitors can walk round quite freely. The chimneypieces of local marble are beautifully worked, and there is an elaborately carved staircase and some fine woodwork. The spacious pleasure grounds – 120 acres of park and 22 acres of garden – hold numerous attractions, including a herd of fallow deer established in the 18th century and a more recently introduced herd of red deer; the Lakeland Motor Museum; the Crafts and Countryside Museum; a baby animal house and an adventure playground. The garden contains rare trees and shrubs as well as the oldest monkey-puzzle tree in England, grown from the original seed first brought into the country by Joseph Paxton in the mid-19th century.

☎ Flookburgh (044853) 328

6½ m S of Haverthwaite on B5278

SD 3577 (OS 97)

Open Easter Su to end Oct daily exc S and Good Fri 1030-1800; reduced rates for parties

⊖ (1 m) P WC ♿ (limited access) 🚌 (by appt)
D 🍴 🍷 🍵 ◆ ✿ 🎣 (by appt) 🚶 ● ⚹

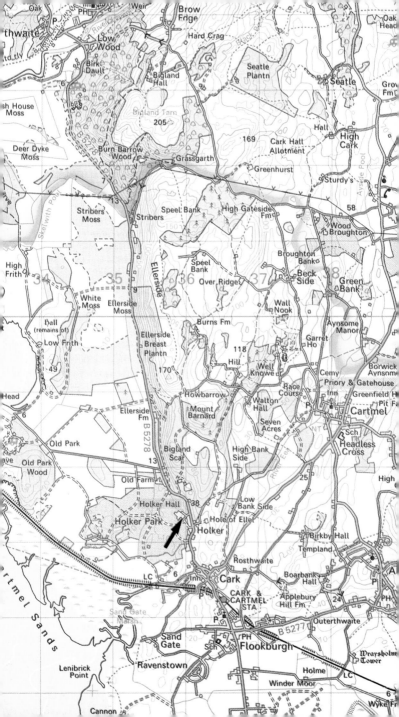

Levens Hall

Kendal, Cumbria

Grey, gabled, irregular and mainly Elizabethan, Levens, like many houses in this region, began life as a pele tower (a square fortified tower house). The base of the tower can still be seen in the undercroft, now the shop. But the façade and most of the rest of the house date from about 1580 when it came into the hands of the Bellingham family, who were also responsible for the house's most outstanding feature, the lavishly decorated series of rooms on the ground floor. The house has been little altered since that time, although a south front was added about a century later by a new owner, Colonel James Grahme, who also made the garden and acquired some fine furniture. The entrance hall leads straight into the great hall, which has a deep plaster frieze decorated with the coat-of-arms of Elizabeth I, animals and heraldic shields. The other ground-floor rooms are beautiful, and most have panelling, carved chimneypieces and moulded plaster ceilings, though in the dining room the walls are covered with Spanish leather, introduced by Colonel Grahme, who probably also bought the fine Stuart chairs. The drawing room, dated 1595, is a particularly rich example of late Elizabethan taste, with heraldic glass, geometric plaster ceilings and an amazing chimneypiece incorporating panels of heraldry and the Orders of Classical architecture. The topiary gardens were laid out in 1690 by Monsieur Beaumont, gardener to James II at Hampton Court. They fortunately escaped the 18th-century craze for landscaping, and remain much as they were planned.

☎ Sedgwick (0448) 60321

6 m S of Kendal on A6 just after junction with A590

SO 4985 (OS 97)

Open Easter Su to early Oct Su-Th 1100-1700; evening parties by appt

⊖ 🅿 WC 🖰 ♣ 🍴 🎋 ◆ ⚞ 𝑘 (by appt for parties) ● (not in house) 🏌

256

Burton Agnes Hall

Great Driffield, Humberside

The lovely proportions of this fine red-brick Jacobean house suggest the hand of a professional designer, and the architect was indeed Robert Smythson, builder of Longleat, Wollaton and Hardwick Hall. Burton Agnes was designed for Sir Henry Griffiths and built between 1601 and 1610. The house is remarkable in that it has been so little altered, though sash windows have replaced the mullioned ones in places, and those at each end of the gallery are 18th century. The brick gatehouse, built in 1610 and bearing the arms of James I flanked by allegorical figures, is a particularly fine example of Tudor architecture. The entrance door, set in the side of a bay, leads into a screens passage and then into the great hall, which contains some truly breathtaking Elizabethan allegorical carving, plasterwork and panelling as well as a fine Nonsuch chest. There is more woodcarving in the drawing room, the centrepiece being a somewhat gruesome Dance of Death over the fireplace. The other rooms were renovated in the 1730s, and reflect mainly 18th-century taste, but the recently redecorated dining room has a splendid carved Elizabethan chimneypiece which was originally in the long gallery, and the staircase, one of the most remarkable features of the house, has been untouched since Tudor times. Like most country houses, Burton Agnes contains some fine paintings, and there is an impressive collection of Impressionist and later paintings. These are on display in the house, many in the long gallery, recently restored and with a lovely barrel ceiling.

☎ Burton Agnes (026289) 324

6½ m SW of Bridlington on A166 at Burton Agnes

TA 1063 (OS 101)

Open 1 Apr to end Oct daily 1100-1700; subject to closure at short notice

♿ 🅿 WC ⬇ (limited access) 🚌 (by appt)
D (grounds) 🍴 🍼 ◆ ⚘ ✗ (by appt) ●

Burton Constable

Sproatley, near Hull, Humberside

A large brick Tudor house with two battlemented towers rising above the roofline, it was mainly built around 1600 for the Constable family, on an earlier house. During the 18th century the main door was moved to the centre and a new top storey added, as well as two bay windows. The interior of the house is the real surprise, as nearly all the interiors are 18th century and very opulent. A large number of architects and craftsmen worked here replacing the original decorations with their own work, and the numbers were added to during the 19th century, when Lady Constable carried out redecorations. The long gallery was made from a series of bedrooms after 1740 with plaster ceilings copied from the Bodleian Library in Oxford; the staircase was built by Timothy Lightholer, who also designed the plasterwork in the hall, the stable block and the new top storey; the lovely ballroom was designed by the young James Wyatt, and the blue drawing room, with its rich Victorian upholstery, was made by Thomas Atkinson in the 19th century. Designs were commissioned but not used from both Robert Adam and Carr of York, and Capability Brown's ceiling design for the great hall was later used at Corsham Court. The furniture is as rich and diverse as the rooms, and there are some good paintings collected by William Constable. The park with its two lakes was laid out by Capability Brown's pupil Thomas White, though Brown advised on it.

☎ Skirlaugh (0401) 62400
9 m NE of Kingston upon Hull on B1238 turn N after Sproatley to Burton Constable
TA 1836 (OS 107)

Open Easter Su to last Su in Sept daily exc F and S 1330-1700

♿ (1½ miles) 🅿 WC 🚻 (limited, by appt)
🍴 (by appt) ♣ ♟ ⛺ ◆ ⚒ 🏄 (by appt) ● 🐕

Normanby Hall

Scunthorpe, Humberside

This fine Regency building, set in a 350-acre park, was built for the Sheffield family, formerly Dukes of Buckingham (and owners of Buckingham Palace). It was designed by Robert Smirke to replace an older house, and was completed about 1830. The rear wing, in the Baroque style, was added by Walter Brierley in 1906, as part of a major programme of enlargement including an east wing with a ballroom and complete range of bedrooms. Much of Brierley's work, however, was demolished after the last war, and the Sheffields themselves left the house in 1963. Since then it has been leased to Scunthorpe Borough Council, and has been completely refurnished in the style of the 1820s. Although it is run as a museum – it contains fine collections of furniture, textiles, uniforms, costumes, paintings, ceramics and silver – the richly decorated rooms have a pleasant 'lived-in' feeling, and there are no rope barriers. The park, which includes a deer park with free-roaming herds of red and fallow deer, has numerous attractions including a golf course, a swimming pool and facilities for horse riding. There are four nature trails, a large wooded picnic area and a centre for traditional crafts, where visitors can watch the potter or the blacksmith at work.

☎ Scunthorpe (0724) 843533/862141

5 m N of Scunthorpe off B1430 at Normanby

SE 8816 (OS 112)

Open Apr to Oct M, W-S 1000-1730, Su pm; Nov to Mar M-F 1000-1700, Su pm; closed 1230-1400

♿ P WC ♿ (limited access) ♨ (by appt) ♠ ☕
🏠 ◆ ♨ 🗡 (evenings only, by appt) 🐕 🐕

Sledmere House

Great Driffield, Humberside

A large 18th-century house standing in a park with great trees, laid out by Capability Brown in 1777. The Sykes family inherited the property in 1748, and the whole ensemble, house, park and village, has been created by them. The village originally lay at the bottom of the valley, but was removed to dry ground out of sight of the house. The square core of the house was built by Richard Sykes, and a generation later Sir Christopher improved it by adding two wings and a pediment. The exterior was finished by 1786, and a year later the interior was decorated with plaster ornament in the Adam style. The plasterer was the famous Joseph Rose, who had worked with Adam on several major houses. Tragically, the house was gutted by fire in 1911, but all the furniture and sculpture and many of the fittings were rescued by the villagers, and the plasterwork was brilliantly restored by the architect Walter Brierley. The staircase hall, drawing room and dining room are all on a grand scale and very fine, and the contents are altogether of a very high quality. Much of the furniture was made for the house, and there are some good paintings, including Elizabethan portraits and 17th-century Italian works. The finest room of all is the library on the first floor, which takes up the whole of the park front. Sir Tatton, son of Sir Christopher, whose collection of books the room was built to house, used to take his exercise here on rainy days, covering miles by walking up and down. The room has a great arched vault based on the baths of ancient Rome, and a lovely parquet floor of 1911.

☎ Driffield (0377) 86208

12 m SE of Malton on B1248 turn NE on B1251 to Sledmere

SE 9364 (OS 101)

Open Apr Easter weekend and Su; May to end Sept daily exc M and F (but inc Bank Hol M) 1330-1730

P WC ♿ (preferably by appt) ♐ (by appt) D (grounds only) ♣ ♥ ♦ ⚹ ⚔ (by appt) ●

Hoghton Tower

Hoghton, Blackburn, Lancashire

The name means 'the house on the hill', and Hoghton, visible for miles in every direction, presents a dramatic sight with its castellated front and crowded chimneys silhouetted on the skyline. The Hoghton family had lived in a house on the site since at least the 12th century, and when Thomas Hoghton began to build his new house in 1565 he built it in a rather old-fashioned medieval style; it was completed by his son in the following century. The house was not lived in during the 18th century as the family moved out in 1710, so it was spared the usual 18th-century rebuilding. However, it did become fairly dilapidated, and the Hoghtons, who moved back again at the end of the 19th century, had to carry out extensive restorations to render it habitable again. The house has two courtyards: the battlemented gatehouse leads into an outer courtyard formed by the stables and service blocks, while the main living rooms are arranged around an inner courtyard. The two were once separated by the great tower that gives the house its name, but this was destroyed during the Civil War. The great hall, with stone walls and an oak roof, contains furniture of its period and relics of James I's visit in 1612 (it was here that he reputedly knighted a piece of beef, thus inventing the word 'Sirloin'). The other rooms are panelled in the 17th-century style. A fire in the 19th century destroyed the best of the furniture and paintings, but there are family portraits, pictures of local interest, and an excellent collection of dolls and dolls' houses.

☎ Hoghton (025485) 2986

4 m W of Blackburn on A675

SD 6226 (OS 103)

Open Easter to end Oct Su, Bank Hol M also S in July and Aug 1400-1800; other times by appt

⊖ P WC ♿ (limited access) ⊟ D (grounds only) ♣ ♥ ⊓ ◆ ⚹ 𝄂 (compulsory) 🕈 ●

Leighton Hall

Carnforth, Lancashire

No finer setting for a house could be imagined: the Lake District mountains
rise blue behind, the park in front slopes gently down to the house, and the
building itself, in brilliant white limestone, shines in the sun. The house was
given its fairy-tale castellated look in the early 1800s; previously it had been a
Classical-style house with a sober façade. The additions were made by
Harrison of Chester, who built turrets and pinnacles and gave the stables a
tall, pointed window so that they look like a chapel from the outside. A
projecting wing with a high tower was added in 1870 by Paley and Austin,
the best architects in the north-west at the time. The house was bought in
1822 by Richard Gillow, whose father had founded the firm of Gillow of
Lancaster; one of its main features is the large amount of very fine furniture
made by this firm. Although Gillows are perhaps best known today for their
19th-century work, they were in business by 1730, and there are some
excellent Georgian pieces in the house, including an altar in the small private
chapel. The interior generally is attractive without being out of the ordinary.
The entrance hall has a delicate screen of clustered columns dividing it from
the elegant curved stone staircase; the dining room, once a billiards room,
still has its central skylight, and the drawing room gives a breathtaking view
of the mountains. The gardens and grounds are both open to visitors, and
birds of prey are flown in the park whenever weather permits.

☎ Carnforth (0524) 734474
10½ m N of Lancaster on A6 turn W to Yealand
Conyers
SD 4974 (OS 97)

Open 1st May to 30th Sept Su, Bank Hol M, T-F
1400-1700

🅵 WC 🅰 🖬 (by appt) D (grounds) ♠ ♥ ⊼ ◆
☘ 𝍒 (compulsory) Display of birds of prey

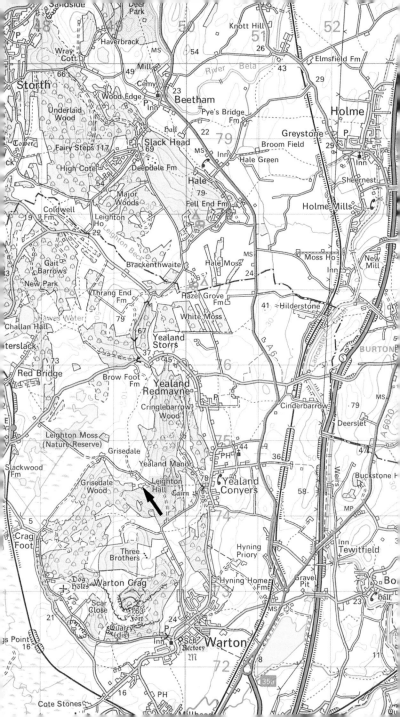

Meols Hall

Southport, Lancashire

This is a brick manor house that is essentially late 17th century in character, though its history goes back much further. There was certainly a house on the site during the reign of King John, when most of the manor of North Meols was granted to Robert de Coudray. It passed by marriage from the de Coudrays to the Aughtons in the 14th century, and in the 16th century to the Heskeths. Like many of the Lancashire gentry, the Heskeths remained Catholics after the Reformation, and there are still traces of a 'priest's hole' at Meols; the famous Jesuit missionary St Edmund Campion is said to have used it. Roger Hesketh and his wife were imprisoned for a time after the Jacobite rising of 1692 in Lancashire, but there nevertheless seems to have been much new building at about this time: an inventory made in 1675 suggests a markedly different arrangement of rooms from the present one, and a back gable on the house, carrying the date 1695, also indicates that changes substantial enough to warrant recording must have been made. In the 18th century the Heskeths married into the Fleetwood family, and Meols was rather neglected in favour of the Fleetwoods' great Rossall estate. But in the 1840s, after the head of the family, Sir Peter Fleetwood-Hesketh, lost most of his fortune, his younger brother Charles managed to save Meols from the wreck. The large Rossall collection (paintings, furniture and *objets d'art*) eventually came to Meols and is the principal attraction of the interior.

☎ Southport (0704) 28171

1 m N of Southport off A527

SD 3618 (OS 108)

Open Apr to end Sept Th and Bank Hol M 1400-1700

♿ 🅿 WC ♿ 🍴 (by appt for parties of more than 20) 🐾 ✗

Dunham Massey

Altrincham, Greater Manchester

Dunham Massey, a low, plain, red-brick house, was mainly an 18th-century rebuilding of an Elizabethan moated manor house built by George Booth, close to the site of a Norman castle. The rebuilding was done by another George Booth, the 2nd Earl of Warrington, using a little-known architect, John Norris. By 1721 he had made extensive alterations to the interiors and laid out much of the park, and in the following years the old Tudor house was completely rebuilt. The arrangement of the new house, however, was dictated by the original Tudor building – today's house still has its open central courtyard, and the great hall is still basically the Tudor hall, though it retains none of its original decorations. Some further alterations were carried out between 1905 and 1909 by the architect J. C. Hall, who added the elaborate stone centrepiece on the south front, and the state rooms were superbly redecorated early in this century under the supervision of the furniture historian Percy Macquoid. Dunham Massey is particularly noted for its fine collections of furniture, silver and paintings, some of which were amassed by the 1st Earl, and others, particularly the pictures, by the 5th Earl of Stamford about 1760. Succeeding generations have added to and cherished the collections. The silver is displayed in Queen Anne's room, the Stamford gallery contains lovely walnut furniture, and there is a collection of paintings of the house in the gallery. The deer park contains an Elizabethan mill in working order, built by the first George Booth.

☎ (061) 941 1025

1½ m W of Altrincham in Dunham Town

SJ 7387 (OS 109)

Open Apr to end Oct daily exc F 1300-1700 (Su and Bank Hol M 1200-1700); reduced rates for parties

♿ (limited) P WC ♿ (limited access) 🚻 (by appt) D (grounds only) ♣ 🍴 ◆ ✲ ● NT

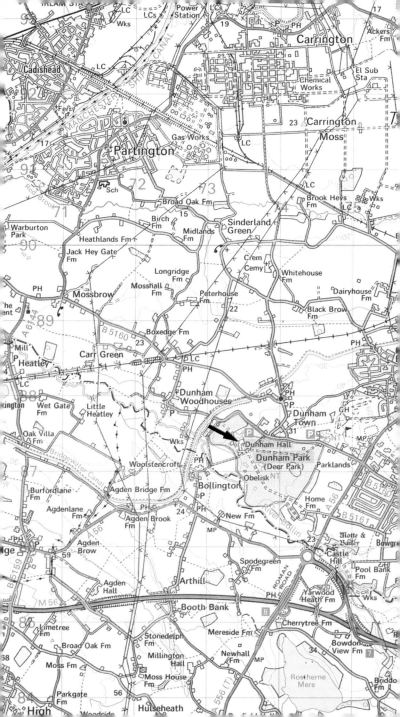

Heaton Hall

Heaton Park, Prestwich, Manchester

Heaton Hall (or House as it was formerly known) is the masterpiece of James Wyatt, one of the greatest architects of the 18th century. His client was Sir Thomas Egerton, later 1st Earl of Wilton, who had inherited the house in 1756 while still a minor. In 1771 Wyatt burst upon the fashionable world with his simplified version of Adam's Neo-Classicism, and Sir Thomas, a man of taste and fashion, commissioned him to prepare designs for the rebuilding of Heaton Hall. The garden front is one of the best examples in the country of the style, and the interior decoration is also very charming, a 'prettier' version of Adam's manner, which was becoming less popular. In 1902, after 200 years of ownership, the Egerton family sold the house to Manchester City Corporation. This was long before the concept of country-house visiting had been born and the entire contents were auctioned off and the house itself used for a variety of unsuitable purposes. However, since 1972 much restoration work has been carried out, and the house is now used as a museum, with good 18th-century paintings and furniture in many of the rooms. A fine central staircase leads to the first floor, where there is a bow-ended room exquisitely decorated in the Etruscan style by Biagio Rebecca. This originally contained gilt and painted chairs and sofas, window seats and a circular carpet.

☎ (061) 236 9422

5½ m N of Manchester city centre on A576 turn W to Heaton Park

SD 8304 (OS 109)

Open early Apr to late Sept M, W-S 1000-1800, Su 1400-1800

⊖ 🅿 WC 🅗 (limited access) 🚻 D (on lead, grounds only) ♣ 🍽 🎪 ★ ◆ ⚘ 🎿 ●

Croxteth Hall

Croxteth Country Park, Liverpool, Merseyside

Until recently the home of the Molyneux family, Earls of Sefton, Croxteth is now under the administration of Merseyside County Museums, and the 500-acre estate inside the city boundaries is organised as a country park. The house was begun in about 1575 for Richard Molyneux, but only a small part of the original Elizabethan house remains, as there were several later and larger additions. The first of these, built between 1702 and 1714, and known as the Queen Anne wing, is today the house's most attractive façade. Raised on a wide terrace and embellished with decorative stonework incorporating the Molyneux arms, it transformed the house into an imposing mansion. Further extensions were made in about 1800, and there were Victorian and Edwardian additions, the north range dating from the 1870s and the west front from 1902. In 1952 the finest rooms in the Queen Anne wing were destroyed by fire, and most have not been restored. There are relatively few outstanding rooms inside, partly because of the fire and partly because most of the contents were sold at auction, but the ground floor has been arranged as the 'Croxteth Heritage' exhibition, which explains the history of the park and the lifestyle of an Edwardian estate. Furniture and other items are being collected for the house, and the rooms are partly furnished, with some of the paintings which originally hung there back on the walls and others on loan from the Walker Art Gallery. Special exhibitions of costume are held from time to time in the main rooms.

☎ Liverpool (051) 228 5311

7 m NE of Liverpool on A580 turn S to Croxteth Country Park

SJ 4094 (OS 108)

Open 6 Apr to end Sept daily 1100-1700; winter hours on request

⊖ P WC ♿ (limited access) ⊟ (by appt) D (not in house) ♣ ☕ ◆ ⁂ 🚶 ● (no tripods) ♿

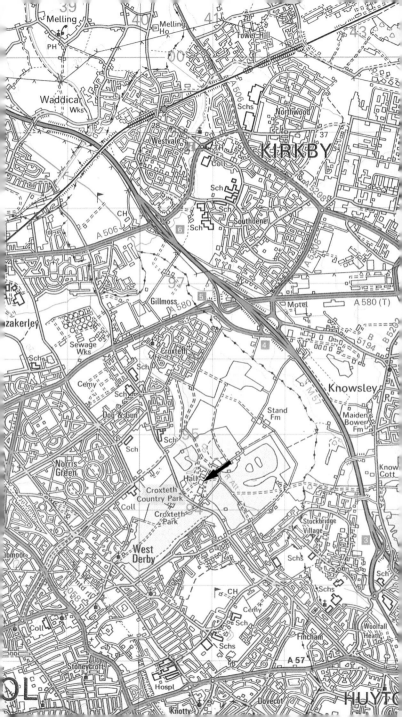

Speke Hall

The Walk, Liverpool, Merseyside

Sandwiched between Liverpool's industrial buildings and a runway of the airport, Speke Hall is an amazing survival – an unspoilt, picturesque Tudor mansion set in its own gardens and woodlands. Built for the Norris family between about 1490 and 1613, it has been remarkably little altered, and the buildings, although of different dates, are all of the same style, the black-and-white half-timbering typical of Cheshire and Lancashire. The house has four wings surrounding a cobbled courtyard in which there are two yew trees older than the house. The great hall is the oldest part of the house, though no one is quite sure of its date. It has a huge battlemented Tudor fireplace at one end, while at the other is sumptuous Flemish panelling containing carved busts of Roman emperors. The parlour has an odd overmantel with a relief representing three generations of the Norris family, and a rich Elizabethan plaster ceiling. The house contains a number of secret hiding places and priest holes, and in most of the rooms there is good oak furniture amassed by the Watt family. Richard Watt bought the house in 1796, and he and his family restored it lovingly, most of the furniture and fittings introduced in the 19th century being in the various Victorian revival styles. Miss Adelaide Watt bequeathed it to descendants of the Norrises, the original owners, and the National Trust received it in 1944. It is now leased to Liverpool Corporation as an historic museum.

☎ Liverpool (051) 427 7231

10 m SE of Liverpool on A561 turn S to Speke Airport

SJ 4182 (OS 108)

Open Apr to Oct T-S 1300-1730, Su and Bank Hol M 1200-1800; Nov to mid Dec S, Su 1300-1700

⊖ (1 m) 🅿 WC 🖺 (limited access) 🚻 (by appt)
♣ 🍴 ◆ ✁ ⚡ ● NT

Carlton Towers

Carlton, North Yorkshire

At first sight this is an uncompromisingly Victorian house, with its amazing array of towers, turrets, pinnacles and gargoyles; but beneath its 19th-century veneer Carlton Towers is very much older. The home of the Dukes of Norfolk, it has passed down by inheritance since the Norman Conquest, and may even still retain some of the medieval masonry, while the Jacobean house, Carlton Hall, can be seen to the left of the entrance. The two men responsible for the house's present appearance were Lord Beaumont and his architect E. W. Pugin, son of the more famous A. W. N. Pugin. Pugin remodelled and refaced it in the 1870s, but he and Lord Beaumont, both eccentrics, then quarrelled, and Pugin was replaced by John Francis Bentley, designer of Westminster Cathedral, who was responsible for most of the interiors, which remain more or less unchanged. His rooms, with their dark, rich colour schemes, took fifteen years to complete, and he designed every detail, right down to such things as curtains, towel rails and firedogs. The most sumptuous room is the pink, green and gold Venetian drawing room, one of the most complete Victorian interiors in existence. The dado panels are painted with scenes from the *Merchant of Venice* by Westlake, the upper parts are covered with moulded and gilded plaster, and the vast chimney-piece is decorated with heraldry and painted panels, again by Westlake. In contrast, several smaller rooms, which retain their 18th-century and Edwardian decoration, give the house a lighter and more 'lived-in' feeling.

☎ Goole (0405) 861662

1 m N of Snaith on A1041

SE 6423 (OS 105)

Open May to end Sept Su and Bank Hol weekends 1300–1700; other times by appt

Castle Howard

Malton, North Yorkshire

From the high ground a few miles west of Malton the mellowed, creamy mass of Castle Howard dominates the countryside. The estate has been in the possession of the Howard family since 1577. The new castle was built between 1700 and 1726 by the 3rd Earl of Carlisle, who chose John Vanbrugh, the soldier turned dramatist, as his architect. Nicholas Hawksmoor assisted him, translating Vanbrugh's grand design into working plans. The design centred on a domed hall inside a monumental entrance. Beyond the hall a long wing, single-storeyed over a basement, contained the principal apartments, while on either side of the entrance a wing was to extend forwards to form a large courtyard. The west wing was completed later than the central block and the east wing, and in a restrained Palladian style in contrast to the exuberant Baroque of Vanbrugh. The house contains collections of furniture, ceramics, and Dutch and Italian paintings, and also the largest collection of costume in private hands. The 19th-century chapel, full of Victorian colour, has some stained glass made by William Morris to designs of Burne-Jones. The 3rd Earl laid out the gardens and park to give good views of the house and to enhance its setting; the landscaping of the park still bears the stamp of the 18th century, though the formal gardens were later redesigned. Vanbrugh's Temple of the Four Winds and the Mausoleum by Hawksmoor are two of the park's features that remain of exceptional interest.

☎ Coneysthorpe (065 384) 333

11½ m NE of York on A64 turn N at Barton Hill for Welburn

SE 7170 (OS 100)

Open 25 Mar to 31 Oct daily 1100-1630; reduced rates for parties

⊖ (limited) ⚑ WC 🅿 🚻 D (on lead, grounds only) ♣ ☘ ◆ ⊞ ✄ ● (not in house)

Newburgh Priory

Coxwold, North Yorkshire

This large stone house is a mixture of styles, Tudor, Jacobean and 18th century, with little remaining of the original Augustinian priory founded in 1145. After the Dissolution Henry VIII sold the buildings to his loyal chaplain Anthony Bellasis, and he began to turn the priory into a mansion, a process that continued until the 18th century. Only some parts of the building can be accurately dated: the porch and mullioned windows are Elizabethan; parts of the courtyard were remodelled in 1720, and the drainpipes on the east front are dated 1732. The ground floor rooms are shown to visitors, and one room upstairs containing Oliver Cromwell's tomb. Thomas Bellasis, the 2nd Viscount Fauconberg, married Cromwell's daughter Mary, but no one really knows why – or even if – the Protector's body was brought here for burial. One of the most notable features of the rooms is the elaborate carved overmantel in the dining room, dated 1615 and made by Nicholas Stone in a style advanced for its time. The black gallery contains good family portraits, including one of Cromwell, and there is a charming 18th-century painting of Earl Fauconberg and his family in the justice room (where the Court Leet was held). More family portraits are hung throughout the rooms and on the stairs, among them two of Mary Cromwell, one showing her in her court robes after her marriage, holding a plan of the drawing room. The two panelled rooms on the garden front were rebuilt after a fire in 1757, and have pretty plaster ceilings and a set of Georgian chairs with needlework covers.

☎ Coxwold (034 76) 435

3½ m SE of Thirsk on A19 turn E to Coxwold

SE 5476 (OS 100)

Open mid May to end Aug W and Su in Aug 1400-1800; parties of 20 or over on other days by appt

P WC 🔊 (limited access) 🚻 D (grounds only)
♣ 🍴 ⚔ 🐕 ● (not in house)

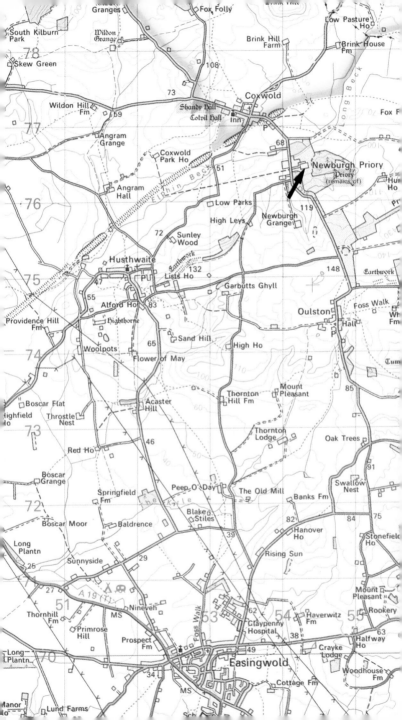

Newby Hall

Ripon, North Yorkshire

This large, red-brick house built in the 1690s for Sir Edward Blackett, was bought in 1748 by the Weddell family. William Weddell, a man of great taste and knowledge, returned from the Grand Tour in the 1760s, determined to alter and enlarge the house for his collection of sculpture and tapestries, and he commissioned the most fashionable architect of the time, Robert Adam. Adam added the two wings to the east of the house, one containing the sculpture gallery and the other the kitchens, and completely redesigned the interiors of the main rooms. Around 1800, Lord Grantham, an amateur architect, turned Adam's dining room into a library and added the 'Georgian dining room' on the north-west corner, and his grandson built the Victorian wing and the billiards room above the dining room. The interiors at Newby Hall provide one of the finest complete examples of the Adam style to be found anywhere, the climax being the great domed sculpture gallery with delicate stucco work in the 'antique' style. This still contains the Roman marbles it was built to display, and Weddell's other collections are also intact – the Gobelin tapestries in the tapestry room were ordered by him in Paris in 1765. The superb furniture was made for the house by Thomas Chippendale, much of it to Adam's own design. The charming bedrooms, several of which have been skilfully refurbished in recent years, also contain fine furniture and decorations. On the back stairs there is a collection of chamber-pots, some very rare, from all over Europe and the Far East.

☎ Boroughbridge (090 12) 2583
2½ m SE of Ripon on B6265 turn S before Bridge Hewick to Great Givendale
SE 3467 (OS 99)

House open 1st Apr to 4th Oct daily exc M but inc Bank Hol M 1200-1700; gardens 1st Apr to 25 Oct

⊖ (1 m) 🅿 WC 🚹 🔒 🌳 ♥ 🍴 ◆ ☼
● (not in house) 🚶 🏃

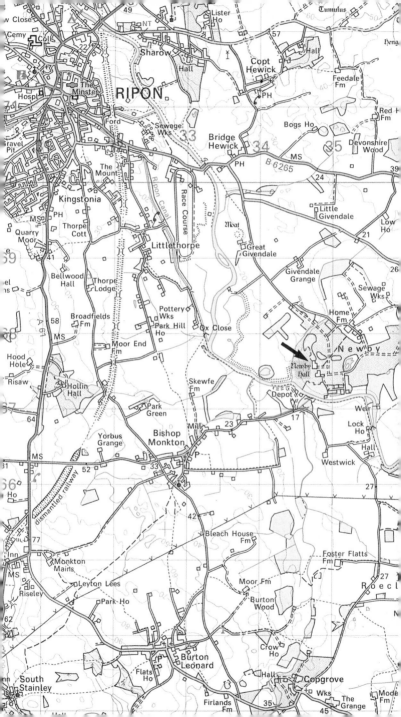

Brontë Parsonage

Haworth, Keighley, West Yorkshire

This rather bleak Georgian house, built in 1778, was the home of the Brontë family, and although the north wing was added in the 1870s, the rest of the house is just as it was in their time. The building was presented to the Brontë Society in 1928, and the north wing now houses the library and exhibition room, while the older part of the house contains the Brontës' furniture and personal possessions. Charlotte made various alterations to the house, enlarging the dining room and bedroom above it; the dining room was where the sisters did most of their work. The room contains the rocking chair where Anne used to sit, the sofa on which Emily died, and a portrait of Charlotte's hero, the Duke of Wellington. Another of Charlotte's alterations was the creation of a study for her fiancé Mr Nicholls, and she wrote to a friend about 'the new little room' with its green and white curtains. Sadly, she survived only a year of married life. The bedroom in which the girls' mother died was also the room in which 'Aunt Branwell' taught the sisters fine needlework, and there are examples on show, together with family china, Charlotte's baby shoes and other items. Some of their toys, which were found under floorboards during repair work, are displayed in the nursery, and 'Branwell's room' contains some of his drawings and paintings. The Bonnell room houses the extensive collection of books, manuscripts and drawings by the Brontës left to the Brontë Society by Henry Houston Bonnell of Philadelphia.

☎ Haworth (0535) 42323

3½ m S of Keighley on A6033, turn W to Haworth

SE 0237 (OS 104)

Open Apr to end Sept daily 1100-1730, Oct to end March daily 1100-1630; closed 24-26 Dec, 1-21 Feb

♿ 🅿 (limited access) 🚽 (by appt) ♣ ◆ ⚐
● (by permission)

Harewood House

Harewood, Leeds, West Yorkshire

Harewood, a showpiece of all that is best in 18th-century architecture and decoration, was built between 1759 and 1771 to the designs of John Carr of York, who also built the stable block and the village. In the 1760s Robert Adam returned from Italy full of ideas and suggestions for improvements, and was given the responsibility for all the interiors, and in 1772 Capability Brown began to landscape the park. The Palladian perfection of the house was slightly marred in the 19th century when Charles Barry built additional floors, modified the south front and removed some of Adam's interior work, but the house is still very impressive, and the interiors are among Adam's finest. The hall shows his mature style, with great Doric half-columns and Neo-Classical plasterwork, while the other rooms are lighter in style, with delicate plasterwork by Joseph Rose and paintings by Angelica Kauffmann. The long gallery, with a lofty ceiling decorated with mythological paintings, was designed for pictures, and contains striking portraits by Reynolds, Gainsborough, Romney and others as well as a collection of Chinese porcelain. Harewood is particularly famous for its superb furniture, most of which was made for the house by Thomas Chippendale, who also designed and made the pelments in the library, carved and painted to resemble drapery. The house also contains a good collection of Italian – mainly Venetian – paintings, and there are several delightful watercolours of the house and park by Turner, Girtin and John Varley.

☎ Harewood (0532) 886225
10½ m S of Harrogate on A61, turn W at junction with A659
SE 3144 (OS 104)

Open 1 Apr to end Oct daily 1100-1615 or 1715; Feb, March, Nov Su only 1100-1515

⊖ (1 mile) P WC 🖼 🚻 (by appt) D (on lead, grounds only) ♣ 🍴 ⛲ ◆ 🎪 ⚘ ☆ 🎣 👤

283

Temple Newsam

Leeds, West Yorkshire

The original Tudor house, four blocks round a courtyard, was built by Thomas, Lord Darcy, early in the 16th century. It was constructed in brick, with stone quoins and mullioned windows. It was rebuilt between 1622 and 1637 in Jacobean style by Sir Arthur Ingram. The east wing was demolished and replaced by a wall and gateway; the north and south wings were rebuilt, and the Tudor west wing was extended at each end to link up with them. Brick walls and mullioned windows and transomed windows were designed to blend with the earlier work so that superficially the whole house looked like a Tudor mansion. Stone balustrades round the roofs of the three wings held the building together. Many alterations have been made at different times to the interior, so that it no longer resembles a 17th-century mansion. The exterior has also been reorganised. The Ingrams continued to own the estate for nearly three hundred years. The last private owner sold the house and 900 acres of parkland to the City of Leeds in 1922. Most of the contents of the house were dispersed at that time, but a nucleus of family portraits and furniture remained, and they have since been supplemented not only by articles of furniture but also by copies of original wallpapers and hangings, contemporary paintings, ceramics and silverware, to show the development of styles during the long occupation of the house. There are now 30 rooms open to the public. Perhaps the most impressive representative of the restoration work is the beautiful Palladian library.

☎ Leeds (0532) 641358

4½ m E of Leeds on A63, turn S at Malton

SE 3532 (OS 104)

Open throughout year Tu-Su and Bank Hol M 1030-1815; May to Sept W 1030-2030 exc 25, 26 Dec

♿ P WC 👓 (limited access) 🚻 D (grounds only) ♣ 🍴 ☂ ◆ 🚼 ✄ ● (by permission) 🚹 🚺

Cleveland

Ormesby Hall, near Middlesbrough, Cleveland (tel Middlesbrough [0642] 324188). 3 m SE of Middlesbrough on A171, turn NE onto B1380. NZ 5216 (OS 93). Mid-18th century house showing the typical lifestyle of the country gentry. The house includes a 17th-century wing; the main house includes fine plasterwork, and a display of 18th-century costume. The 18th-century stable-block is notable. Open W, Th, S, Su and Bank Hols pm, Apr to Oct; parties at other times by appt. ⊖ 🅿 WC 🅰 (limited access) 🍴 (by appt) ♣ 🍶 ◆ ⚹ 𝕂 ● (no tripods, flash) NT

Cumbria

Abbot Hall Art Gallery, Kendal, Cumbria (tel Kendal [0539] 22464). In town centre, by the River Kent. SD 5192 (OS 97). House built in 1759 for Col. George Wilson. The ground floor has been restored in 18th-century style, and the upper floor is a gallery for modern art. The stable block contains a museum of Lakeland life and industry. Open daily 1030-1730 (S and Su 1400-1700); closed 25-26th Dec,1st Jan, Good Fri. ⊖ 🅿 WC 🅰 (limited access) 🅰 (by appt) ♣ 🍶 (summer only) ◆ ⚹ 𝕂 (by appt) ⚕ ● (with permission)

Brantwood, Coniston, Cumbria (Coniston [0966] 41396). 6 m SW of Ambleside on B5285, turn SW. SD 3195 (OS 96). Mainly 19th-century house that was the home of John Ruskin from 1872 to 1900, with an exceptional view over Coniston Water. The house has a collection of Ruskin's water-colours and memorabilia. The grounds contain a fine nature trail. Open March to Nov daily am and pm; Nov to March W to Su, am and pm. ⊖ (by boat) 🅿 WC 🅰 🅰 (by appt) D (not in house) ♣ 🍶 ◆ ⚹ 𝕂 (by appt) ● (no flash) ⚕

Castletown House, Rockcliffe, Carlisle, Cumbria (tel Rockcliffe [0228 74] 205). 5 m NW of Carlisle on W side of A74. NY 3462 (OS 85). Georgian country house in attractive grounds. There is a display of naval paintings. Open Apr to Sept W and Bank Hol M 2-5; other times by appt. ⊖ 🅿 WC 🅰 🍶 (for parties by appt) 𝕂

Dalemain, near Penrith, Cumbria (tel Pooley Bridge [085 36] 450). 1 m W of Penrith on A66, turn SW onto A592. NY 4726 (OS 90). Georgian manor house built on the site of an older manor house. The drawing room has hand-painted Chinese wallpaper, and Chinese Chippendale furniture. The fireplace in the drawing room is by Grinling Gibbons. The Norman tower of the old house contans a museum of the local regiment. Open Easter Su to mid-Oct daily exc F and S am and pm. 🅿 WC 🅰 🅰 D (on leads) ♣ 🍶 🅰 ⚹ 𝕂 ● (not in house) ⚕

Dove Cottage, Grasmere, Cumbria (tel Grasmere [096 65] 464). ½ m SE of Grasmere village on A591, turn SE. NY 3407 (OS 90). The home of William Wordsworth from 1799 to 1808, after which it was the home of Thomas de Quincy until 1835. It has been preserved as a Wordsworth memorial since 1890. The house reconstructs the life of the late 18th century and contains a collection of manuscripts. Open daily am and pm, closed mid Jan to mid Feb. ⊖ 🅿 WC 🅰 (limited access) 🅰 (by appt) ♣ ◆ ⚹ 𝕂 ● (not in cottage)

Hill Top, Near Sawrey, Hawkshead, Cumbria (tel Hawkshead [096 66] 269). 3 m S of Hawkshead on B5285, in village of Near Sawrey. SD 3695 (OS 97). 17th-century farmhouse, the home of children's author and illustrator Beatrix Potter until 1913, and retained unaltered since that date. A collection of her illustrations is on display. Open Apr to Oct daily (exc F) 1000-1700 (Su

1400-1700). 🅿 WC ♿ (limited access) 🅗 (by appt) ♣ ◆ ☕ ● (not in house)

Hutton-in-the-Forest, Penrith, Cumbria (tel Skelton [085 34] 207). 3 m N of Penrith on A6 turn W onto B5305 for 5 m. NY 4635 (OS 90). 14th-century pele tower with many later additions; the house was greatly rebuilt in the early 17th century. The interior was remodelled by Anthony Salvin in the 19th century. The gardens date from the 17th century. House open late May to mid-Sept Th, F, Su and Bank Hol M 1300-1600; parties at other times by appt; grounds open daily am and pm. 🅿 WC ♿ (limited access) 🅗 (by appt) ♣ ☕ ◆ ☕ ✗ (by appt) ● (by permission) ♨

Rydal Mount, Ambleside, Cumbria (tel Ambleside [096 63] 3002). 3 m N of Ambleside on A591, turn N. NY 3606 (OS 90). The home of William Wordsworth from 1813 to 1850, containing a collection of books and personal effects. The garden was laid out by Wordsworth himself and is particularly attractive. Open daily am and pm (exc T, Nov to Mar). ⊖ WC ♿ (limited access) 🅗 D (not in house) ♣ ◆ ☕ ● (not in house)

Swarthmoor Hall, near Ulverston, Cumbria (tel Ulverston [0229] 53204). 1 m SW of Ulverston on A590, turn E. SD 2877 (OS 96). Early 17th-century hall built by George Fell and with early associations with the Quakers. The interior contains 17th-century furniture and relics of the early history of the Society of Friends. George Fox, who married Fell's widow Margaret, built the Meeting House here in 1668. Open March to Oct M, T, W, S am and pm; Th by appt. ⊖ (1 mile) 🅿 WC ♿ (limited access) 🅗 (by appt) ★ ☕ ♨

Townend, Troutbeck, Cumbria (tel Ambleside [096 63] 2628). 3 m SE of Ambleside, in Troutbeck village. NY 4002 (OS 90). Small farmhouse built by George Browne in 1626, occupied

by his family until 1944 and typical of Lakeland houses. The interior contains a rich and diverse collec-tion of woodwork carved by the Browne family over the centuries. The dairy houses old farm equipment. Open Apr to Oct daily (exc M and S) pm. ⊖ 🅿 WC 🅗 (by appt) ☕ NT

Wordsworth House, High Street, Cockermouth, Cumbria (tel Cockermouth [0900] 824805). In town centre. NY 1230 (OS 89). Mid-Georgian house where William Wordsworth was born in 1770. The house contains 18th-century furniture, and objects connected with the poet. The garden leads down to the River Derwent. Open Apr to Oct daily (exc Th) am and pm (Su pm only). ⊖ WC 🅗 (by appt) ♣ ☕ ◆ ☕ ● (no flash, no video)

Humberside

Blaydes House, High St, Hull, Humberside (tel Hull [0482] 26406). In town centre. TA 0928 (OS 107). Merchant's house of the mid-Georgian period, with a fine stair-case and panelled rooms, restored by the East Yorkshire Georgian society. Open all year M-F office hours by appt. ⊖ 🅗 (by appt) ☕

Lairgate Hall, Lairgate, Beverley, Humberside (tel Hull [0482] 882255). In town centre. TA 0339 (OS 107). Large house built for the Perryman family in about 1700, now used as council offices. The upstairs room that is now used as the council chamber contains fine 18th-century plasterwork and wallpaper. Open M-F am and pm (office hours). ⊖ 🅿 ♿ ★ ☕

Maister House, High St, Hull, Humberside (tel Hull [0482] 24114). In city centre. TL 1028 (OS 107). Merchant's house built in 1744 in the Palladian manner; its staircase hall is distinctive, decorated by Joseph Page with ironwork by Robert Bakewell. Hall and stairway only open M-F 1000-1600 (closed Bank Hols). ⊖ ⛟ ● NT

Norman Manor House, Burton Agnes, Humberside (tel Burton Agnes [026 289] 324). 6 m SW of Bridlington, on A166. TA 1064 (OS 101). Manor house dating from the Norman period, with original architectural features within its 18th-century façade. Open daily am and pm. ⊖ ▯ ★

Old Rectory, Epworth, Humberside (tel Epworth [0427] 872268). 13 m SW of Scunthorpe. SE 7804 (OS 112). Early home of John and Charles Wesley; restored in 1957, with objects associated with Wesley and the Methodist movement. Open March to Oct daily am and pm (Su pm only). ▯ WC ⛟ (limited access) ⊟ (by appt) ♣ ◆ ⛟ 𝕏 ● (not in house)

Sewerby Hall, Bridlington, Humberside (tel Bridlington [0262] 7855). 2 m NE of Bridlington on B1255, turn S for ½ m. TA 2069 (OS 101). Manor house rebuilt in about 1715 and adapted in the early 19th century. The interior contains old furniture, an art gallery and a collection of Amy Johnson relics. The gardens are of great botanical interest, with an Old English walled garden. There is also a miniature zoo and aviary. House open Spring Bank Hol to mid Sept daily am and pm; park daily am and pm. ⊖ ▯ WC ⛟ ⊟ (by appt) D (on lead) ♣ ● ⊼ ★ (to Hall) ◆ ❀ ⛟ 𝕏 ≀

Wilberforce House, High St, Hull, Humberside (tel Hull [0482] 223111 ext 2737). In city centre. TL 0928 (OS 107). 17th-century town mansion, the birthplace of William Wilberforce in 1759. It contains a local history museum, and a display connected with Wilberforce's campaign for Emancipation. The gardens are also open. Open M-S am and pm, Su pm only. ⊖ ▯ WC ⛟ (limited access) ⊟ (by appt) ♣ ⊼ ★ ◆ ⛟ 𝕏 ≀ (by appt)

Lancashire

Astley Hall, Chorley, Lancashire (tel Chorley [025 72] 62166). 2 m NW of Chorley town centre on A581, turn N onto B5252 for ½ m, then turn E. SD 5718 (OS 108). Elizabethan house with many-windowed façade set in a large park. The interior contains highly elaborate Restoration plasterwork, Cromwellian furniture, Flemish tapestries and a magnificent long gallery. Open Apr to Sept daily 1200-1800; Oct to Mar M-F 1200-1600, S 1000-1600, Su 1100-1600, closed 25-26th Dec, 1st Jan. ⊖ ▯ WC ⛟ ⊟ (by appt) D (on lead, grounds only) ♣ ● ⊼ ◆ ⛟ ● (with permission)

Browsholme Hall, near Clitheroe, Lancashire (tel Stonyhurst [025 486] 330). 5 m NW of Clitheroe off B6243. SD 6845 (OS 103). 16th-century house, with early and late 18th-century additions. It has been the home of the Parker family ever since it was built, and contains a wide-ranging collection of antiquities, furniture and portraits. The grounds were landscaped in the early 19th century. Open Easter weekend, weekend following late Spring Bank Hol S-M; June to Aug S am, pm. ▯ WC ⛟ ⊟ (by appt) ♣ ● (by appt) ⊼ ◆ ⛟ 𝕏

Chingle Hall, Goosnargh, near Preston, Lancashire (tel Broughton [0772] 861082). 5 m N of Preston on A6, turn E onto B5269. SD 5635 (OS 102). Medieval manor house with 13th-century door. It is claimed to be the most haunted house in Britain. Open daily 1000-1800, exc 25 Dec, 1 Jan. ⊖ ▯ WC ⛟ ⊟ ♣ ● ⊼ ◆ ⛟ 𝕏 ≀

Gawthorpe Hall, Padiham, near Burnley, Lancashire (tel Padiham [0282] 78511. 2 m W of Burnely on A671, turn N to E ouskirts of Padiham. SD 8034 (OS 103). Fine early 17th-century manor house retored in the 1850s by sir Charles Barry. It was built around a pele tower, is unusual in having the great hall at the back of the building, and has a minstrel gallery. The house contains the Kay-Shuttleworth collection of textiles and embroidery, mainly of the 19th and early 20th centuries. There is also a collection of early European furniture. Open 1st Apr to 1st Nov daily exc M, F, am and pm; also Good Fri pm. Gardens open all year. Craft gallery open Feb to Dec daily am and pm (Su pm only). ⊖ 🅿 WC 🅱 (limited access) 🚻 (by appt) D ♣ 🐾 ♦ ⚶ 𝕁 NT

Rufford Old Hall, Rumford, near Ormskirk, Lancashire (tel Rufford [0704] 821254). 9 m NE of Ormskirk on A59. SD 4616 (OS 108). Elizabethan half-timbered house with later wings. There is an exceptional 15th-century great hall with carved timber roof and a unique movable screen. The house also contains collections of Tudor arms and armour, 17th-century furniture and the Rufford village museum of social history. Open 1st Apr to 1st Nov daily (exc F) pm. ⊖ 🅿 WC 🚻 (by appt, not Su) 🐾 ♣ 🐾 ♦ ⚶ ● (no flash, tripods, video) NT

Samlesbury Hall, Samlesbury, near Preston, Lancashire (tel Mellor [025 481] 2010). 6 m E of Preston on A677. SD 6230 (OS 102). 14th-century black-and-white timber-framed manor house, partly rebuilt in the 19th century. The hall has an elaborate timber screen of 1532. The house contains exhibitions run by the Council of the Preservation of Rural England, often with antiques for sale. Open T-Su all year, am and pm (closed Christmas to mid Jan). ⊖ 🅿 WC 🅱 (limited access) 🚻 (by appt, not weekends) ♣ 🐾 🚬 ⚶

Turton Tower, Turton, Lancashire (tel Bolton [0204] 852203). 5 m N of Bolton on B639. SD 7315 (OS 102). 15th-century pele tower with an Elizabethan farmhouse attached. There is a display of armour and

local history. The house stands in 8 acres of grounds. House open daily in high summer; gardens open daily throughout year. ⊖ 🅿 WC 🚻 (by appt) ♣ 🐾 (Su only) 🚬 ♦ ⚶ 𝕁

Whalley Abbey, near Blackburn, Lancashire (tel Whalley [0254 82] 2268). 8 m N of Accrington on A680, turn W onto B6246, for ½ m then S. SD 7235 (OS 103).Ruined Cistercian abbey built in the early 14th century; little of the church remains, but some of the monastic buildings survive including the 14th-century gatehouse, and among them is a 16th-century manor house now used as a retreat and conference centre. The grounds stretch down to the river. Open daily am and pm. ⊖ 🅿 🅱 🚻 (by appt) D (on lead) ♣ 🐾 (for parties only) 🚬 ♦ (Easter to Sept) ⚶ 𝕁 (by appt)

Greater Manchester

Fletcher Moss, Wilmslow Rd, Didsbury, Greater Manchester (tel 061-236 9422). 5 m S of city centre on A5145. SJ 8491 (OS 109). Small parsonage built in the early 19th century, now preserved as a local heritage centre, with works of art by artists of the north west. The house is set in an attractive small garden. Open Apr to Sept daily excl T am and pm (Su pm only). ⊖ WC 🚻 (by appt) ♣ 🐾 ★ ● (no flash)

Foxdenton House, Foxdenton Park, Chadderton, Greater Manchester (tel Royton [061-620] 3505). 5 m NE of city centre on B6189. SD 9005. (OS 109) Small mansion of red brick built between 1710 and 1730. The house was restored in the 1960s. Open Apr to Sept T, Su and Bank Hols pm. ⊖ 🅿 WC 🚻 (by appt) ♣ ★

Hall i' th' Wood, Crompton Way, Bolton, Greater Manchester (tel Bolton [0204] 51159. 2 m NE of town centre on A58. SD 7211 (OS 109). Black-and-white timber-framed house from the 1480s, with additions in 1591 and 1648. It was the home of Samuel Crompton, inventor of the cotton spinning machine in 1779. The house was restored in the early 20th century by Lord Leverhulme as a 17th-century home. Open M-W, F-S 1000-1700, Apr to Sept also Su 1400-1715. ⊖ 🅿 WC 🚻 (by appt) ♠ ◆ ☙ ⚔ (by appt) ● (not in house)

Newton Hall, Dukinfield Rd, Hyde, Greater Manchester. 7 m E of city centre, 1½ m N of Hyde on B6170. SJ 9596 (OS 109). 14th-century timber-framed manor house, comprising a single large hall. It has been thoroughly restored, and is set in large open grounds. Open M-F am and pm. ⊖ 🅿 ♿ ♠ ★

Ordsall Hall, Taylorson St, Salford, Greater Manchester (Tel 061-872 0251). 2 m SW of Manchester city centre on A57. SJ 8197 (OS 109). Fine timber-framed house of 16th and 17th centuries, with intact great hall, and fully-equipped Victorian kitchen. On the upper floor is a display of local social history and artefacts. Open M-F, am and pm, Su pm only (closed Christmas, New Year's Day, Good Fri). ⊖ WC ♿ (limited access) 🚻 (by appt) ♠ ★ ☙

Platt Hall, Platt Fields, Rusholme, Greater Manchester (tel 061-224 5217). 3 m SE of city centre off A6010. SJ 8594 (OS 109). Mid 18th-century house built by Timothy Lightoler of red brick. The interior contains a large and important gallery of English costume, with clothing items, accessories and related books from the 17th century to the present day. Open daily am and pm (Su pm only). ⊖ ♿ (limited access) 🚻 ♠ ◆ ● (with permission) ★

Smithills Hall, Smithills Dean Road, Bolton, Greater Manchester (tel Bolton [0204] 41265). 2½ m NW of Bolton centre on A58, turn N. SD

6911 (OS 109). One of Lancashire's oldest manor houses, dating in part from the 14th century, though added to over the years. The great hall is 15th-century, with a fine exposed timber roof. The extensive grounds include a nature trail and associated museum. Open daily am and pm exc Th, Su am (all day Oct to Mar) and Christmas, New Year and Good Fri. ⊖ 🅿 WC ♿ D 🚻 ♠ ☰ ⚔ 🗡 ⚓

Wythenshawe Hall, Northenden, Greater Manchester (tel 061-236 9422). 7 m S of city centre on A60, turn W onto B5167. SJ 8189 (OS 109). Timber-framed manor house with Georgian additions, the home of the Tatton family for many centuries. It houses a wide-ranging collection of *objets d'art*, including paintings, arms and armour, ceramics, ivories and Japanese prints. The house stands in an attractive park. Open Apr to Sept daily exc Tu am and pm (Su pm only). ⊖ 🅿 WC ♿ (limited access) 🚻 (by appt) D (park only) ♠ 🐾 ☰ ★ ☙ 🗡 (by appt) ● (with permission, no flash) ⚓

Merseyside

Bluecoats Chamber, School Lane, Liverpool, Merseyside (tel 051-709 5297). In city centre. SJ 3591. (OS 108). A fine building of the early 18th century with a cobbled quadrangle and a garden courtyard. It was originally a charity school but now serves as a gallery and concert hall. Open T-S 1030-1700; closed Bank Hol weekends, 25 Dec to 1 Jan. ⊖ WC 🚻 ♠ 🐾 ★ ● (with permission)

Northumberland

Belsay Hall, Belsay, Northumberland (tel Belsay [066 181] 636). 14 m NW of Newcastle on A696. NZ 0878 (OS 88). Early 19th-century Neo-Classical house built in the grounds of a 14th-century tower-house castle. Sir Charles Monck designed the Hall in immense detail and planned its gardens. The stable block, servants wing, gardens and castle are all open to the public. Open Apr to Sept daily am and pm; Oct to Mar closed S, Su. 🅿 WC ♿ (limited access) 🚻 D (on lead, grounds only) ♣ 🍴 ◆ ⚘ EH

Cragside House and Country Park, Rothbury, Northumberland (Rothbury [0669] 20333). 1 m E of Rothbury on B6341, turn N. NU 0702 (OS 81). House built by Richard Norman Shaw for Lord Armstrong in the late 19th century; the first house to be lit by water-generated electricity. It contains Arts and Crafts furniture, and is set in a 900-acre country park. House open 1st 2 weeks Apr, Oct, W, S, Su pm; May to Sept daily (closed M, exc Bank Hol M) pm; grounds open Apr to Oct am and pm. ⊖ (limited) 🅿 WC ♿ D (grounds only) 🚻 ♣ 🍴 🛏 ◆ ⚘ ● (no flash in house) ☨ NT

George Stephenson's Birthplace, Wylam-on-Tyne, Northumberland (tel Wylam [06614] 3457). 8 m W of Newcastle on A69 turn S at Wylam. NZ 1265 (OS 88). A small stone-built cottage, built in the mid-18th century and furnished in late 19th-century style. The inventor was born there in 1781. Open Apr to end Oct W, Th, S and Su pm. ⊖ 🅿 (½ mile) ● (no flash) NT

Howick Hall, Howick, Northumberland (tel Longhoughton [066 577] 285). 4 m NE of Alnwick, on B1340; turn onto B1399 for 1 m then turn E. NU 2417 (OS 81). Late 18th-century house with fine grounds laid out by Earl Grey, including many shrubs and rhododendron gardens. Grounds only open Apr to Sept daily, pm. ⊖ 🅿 WC 🚻 D (on lead, grounds only) ♣

Meldon Park, Morpeth, Northumberland (tel Hartburn [067 072] 661). 7 m W of Morpeth on B6343. NZ 1085 (OS 81). Neo-Classical house built in 1832 by John Dobson for the Cookson family. The rooms have

290

spacious views over the country-side, and the 10-acre wooded grounds are famous for their rhododendrons. Open daily from late May Bank Hol Su for 4 weeks. 🅿 WC ♿ 🚻 ♣ 🍴 (weekends only) ⚘ ✗ (by appt) ● (not in house)

Seaton Delaval Hall, Northumberland (tel [091] 2371493). 12 m NE of Newcastle on A190. NZ 3276 (OS 88). Villa built in the 1720s by Vanbrugh for Admiral Delaval near the coast, in a dramatic style combining Classical and Eliza-bethan features. Much of the house was ruined by fire in the 19th century, but part has been restored and houses a collection of paintings. There are fine stables in one wing. Closed until May 1988, then open 1st May to end Sept W, Su and Bank Hol M pm. ⊖ 🅿 WC 🚻 D (on lead) ♣ 🛏 ⚘

Wallington House, Cambo, Northumberland (tel Scots Gap [067 074] 283). 15 m W of Morpeth on B6343, turn S onto B6342. NZ 0284 (OS 81). House built in 1688, in Classical style, modified in the mid-18th century with fine wood- and plasterwork. A central hall, added in the 19th century, was decorated by William Bell Scott and John Ruskin. There is a display of Victorian dolls' houses, and of old coaches. The grounds were laid out by Capability Brown, and include a walled terraced garden, woodlands and lakes. House open mid Apr to Sept daily (exc T) pm; early Apr and Oct W, S, Su pm. Grounds open daily am and pm. ⊖ (limited) 🅿 WC ♿ (limited access) 🚻 (by appt) D (grounds only) ♣ 🍴 🛏 ◆ ⚘ ● (no flash) NT

Tyne & Wear

Washington Old Hall, Washington, Tyne & Wear (tel Washington [091] 4166879). 5 m W of Sunderland on A1231. NZ 3156 (OS 88). Jacobean manor house, with parts dating back to the 12th century. It was the home of the Washington family, of which George Washington was a member. The house has been restored since the 1930s and furnished in 17th-century style. Open Apr and Oct W, S, Su; May to Sept daily exc F am and pm. ⊖ 🅿 WC 🅰 (limited access) 🅿 (by appt) D (on lead, grounds only) ♣ ☞ 🎋 ◆ ☀ ⬤ (no video)

North Yorkshire

Bedale Hall, Bedale, North Yorkshire (tel Bedale [0677] 24604). 10 m SW of Northallerton on A684. SE 2688 (OS 99). Mid-Georgian mansion, with fine plasterwork; the outstanding room is the ballroom, which has a fine decorated ceiling. There is also a collection of items of local interest. Open May to Sept daily 1000-1600; Oct to Apr 1000-1600. ⊖ (limited) 🅿 WC 🅰 🅰 (by appt, T only) ♣ ☞ ★ ⊖ ☀ 𝒌 ♨

Beningbrough Hall, Shipton-by-Beningbrough, North Yorkshire (tel York [0904] 470715). 8 m NW of York on A19, turn W. SE 5158 (OS 105). Attractive red-brick Queen Anne house built in 1716, with carved woodwork by William Thornton. The house is used to display a collection of 100 portraits of the period 1688-1760 belonging to the National Portrait Gallery. The Victorian laundry carries a display of 19th-century domestic life. Open May to Oct T-Th, Bank Hol M pm; Apr, Nov S, Su pm; also daily in Easter week (exc Good Fri) pm. 🅿 WC 🅰 (limited) 🅰 (by appt) ♣ ☞ 🎋 ◆ ☀ 𝒌 (by appt) ⬤ (no flash, tripods) NT

Broughton Hall, near Skipton, North Yorkshire (tel Skipton [0756] 2267). 3½ m W of Skipton on A59. SD 9450 (OS 103). Georgian house built in the Classical style 1750-1810 for the Tempest family. There are many interesting family portraits, and a private chapel. The conservatory dates from 1855, as does

the Italian garden. Open June M-F 1400-1700, and Spring and Summer Bank Hol M, Easter M 1000-1900. ⊖ 🅿 WC 🅰 (by appt) ♣ 𝒌 (compulsory) ⬤ (not in house)

Ebberston Hall, Scarborough, North Yorkshire (tel Scarborough [0723] 85516). 14 m SW of Scarborough on A170. SE 8983 (OS 101). Palladian villa designed by Colen Campbell in 1718 as a pavilion for a large garden that mostly no longer survives. The interior contains some bold wood-carving. Open Easter to Oct daily 1000-1800. ⊖ 🅿 WC 🅰 (limited access) 🅰 (by appt) D (on lead, grounds only) ♣ 🎋 ◆ ☀ 𝒌 (by appt) ⬤ (by permission only)

Kiplin Hall, near Richmond, North Yorkshire (tel Richmond [0748] 818178). 7 m E of Richmond on B6271. SE 2797 (OS 99). Small house with lively skyline, built for the 1st Lord Baltimore in 1625. The interior was redesigned in the early 18th century, when a new staircase was installed. Open May to Sept W and Su 1400-1700. 🅿 WC 🅰 (limited access) 🅰 (by appt) D (on lead, grounds only) ♣ ☞ (Su pm only) 🎋 ☀ 𝒌 (by appt)

Markenfield Hall, Ripon, North Yorkshire. 4 m S of Ripon off A61, turn W. SE 2967 (OS 99). Medieval moated manor house, with a great hall of the early 14th century; the house was abandoned in the 1560s, but has been considerably restored since 1981. Open Apr to Oct M am and pm. 🅿 🅰 (limited access) 🅰 (by appt) 𝒌 (parties only, by appt)

Norton Conyers, Ripon, North Yorkshire (tel Melmerby [076 584] 333). 3½ m N of Ripon on A61, turn N. SE 3176 (OS 99). Attractive house of the early 16th century with an 18th-century façade. It has been the home of the Graham family since 1624. The James II bedroom contains notable 17th-century furniture. There is a walled garden, specialising in 18th-century and unusual hardy plants. Open June to Sept Su; also Easter and May Bank Hol Su and M; and end July to early Aug daily. ⊖ 🅿 WC ♿ (limited access) 🛏 (by appt) 🚌 (parties only, by appt) 🍴 (open all year M-F, also S, Su Apr to Sept)

Nunnington Hall, near Helmsley, North Yorkshire (tel Nunnington [043 95] 283). 14 m SE of Helmsley on B1257, turn E. SE 6779 (OS 100). Large manor house of the 17th century and earlier, rebuilt by Lord Preston after 1688. There is much woodwork of this date, and a fine chimneypiece in the panelled hall. A collection of miniature rooms, fully furnished in the style of various periods, can be found in the attic. Open Apr to Nov S and Su; May to Oct T-Th pm; also Bank Hol M pm. 🅿 WC ♿ (limited access) 🛏 (by appt) 🚌 🍴 ♣ ◆ ᛉ 🗡 (by appt) ● (no flash, tripods) NT

Osgodby Hall, near Thirsk, North Yorkshire (tel Thirsk [0845] 597534). 4 m E of Thirsk on A170, turn S. SE 4980 (OS 100). Small and elegant Jacobean house, with a broad 17th-century staircase. The fine forecourt remains as it was built in 1640. Open Easter to Sept S, Su and Bank Hols pm. 🅿 🛏 (by appt) ♣ 🚌 🗡

Ripley Castle, Ripley, North Yorkshire (tel Harrogate [0423] 770152). 3½ m N of Harrogate, on A61. SE 2860 (OS 99). The home of the Ingilby family for 600 years, the castle is now an 18th-century mansion built around a 15th-century tower. The original gatehouse of the 1410s survives. The gardens are 18th-century and the model cottages of the village were created in the 1820s. The house contains an exhibition of Civil War armour and weapons, and other curiosities. Open Easter to early Oct daily exc M in July and Aug am and pm (gardens daily Easter to Oct). ⊖ 🅿 WC 🛏 (by appt at any time) ♣ 🚌 ᛉ ◆ ᛉ 🗡 🪵

Shandy Hall, Coxwold, North Yorkshire (tel Coxwold [034 76] 465). 4 m SE of Thirsk. SE 5377 (OS 100). Small medieval hall that was the home of Laurence Sterne from 1760 to 1767, the place where he wrote *Tristram Shandy* and *A Sentimental Journey*. It is not kept as a museum to Sterne, but as a 'lived-in house full of relevant books, pictures, memorabilia – and surprises!'. There is an attractive walled garden. Open June to Sept W and Su pm; other times by appt. 🅿 WC 🛏 (by appt) ♣ ᛉ ◆ ᛉ 🗡 ● (not in house)

Stockeld Park, Wetherby, North Yorkshire (tel Wetherby [0937] 66101). 8 m SE of Harrogate on A661, turn SW. SE 3749 (OS 104). Small country mansion, a notable example of the work of James Paine and built between 1758 and 1763. The central hall and staircase are impressive. Open mid-July to mid-Aug daily (exc M) pm. ⊖ 🅿 WC 🛏 ♣ 🚌 ◆ ᛉ 🗡 ● (not in house)

Sutton Park, Sutton-on-the-Forest, North Yorkshire (tel Easingwold [0347] 810249). 12 m N of York on B1363. SE 5864 (OS 100). Early Georgian house of about 1730, built in the Palladian style. There is interesting 18th-century plaster-work, furniture from both England and France, and a collection of porcelain. Open Easter S, Su, M, pm; Apr Su pm; May to early Oct Su, T, Bank Hol M pm. Garden open daily Easter to early Oct. Parties at other times (exc S) by appt. ⊖ 🅿 WC ♿ 🛏 ♣ 🚌 ◆ ᛉ ● (not in house) 🪵

Treasurer's House, Chapter House St, York, North Yorkshire (tel York [0904] 24247). In city centre, to N of Minster. SE 6052 (OS 105). Large house of the 1630s, built for the treasurer of the Minster. It was restored in 1900, but the interior retains many 18th-century features. The house contains period furniture, paintings and a display of its historical associations. There is a small formal garden. Open Apr to Oct daily (exc Good Fri) am and pm. ⊖ WC ♿ (limited access) ☕ ◆ ✱ 🚶 (by appt) ● (no flash, tripods) ⊟ ♣ NT

South Yorkshire

Cannon Hall, Cawthorne, South Yorkshire (tel Barnsley [0226] 790270). 6 m W of Barnsley on A635, turn NW. SE 2708 (OS 110). Modest house, mainly built in the 1760s by John Carr of York, with a Jacobean-style ballroom of the 1890s. It is now a museum, housing 18th-century furniture, glassware, paintings of the Dutch Old Masters, and the regimental museum of the history of the 13th/18th Royal Hussars. There is a large park. Open daily M-S 1030-1700, Su 1430-1700; closed 25-27th Dec, Good Fri. ⊖ P WC ♿ (limited access) ⊟ (by appt) D (on lead, grounds only) ♣ ★ ✱

Cusworth Hall, Doncaster, South Yorkshire (tel Doncaster [0302] 782342). 2 m NW of town centre on A638, turn SW. SE 5403 (OS 111). Mid 18th-century house built for the Wrightson family, with wings added by James Paine. There are fine plaster ceilings and chimney-pieces, and the house is used as a museum of South Yorkshire life. There is a large park with many facilities. Open M-Th, S 1100-1700; Su 1300-1700 (Nov to Feb closes at 1600); closed 25, 26 Dec. ⊖ (½ mile) P WC ♿ (limited access) ⊟ (by appt, Mar to Oct) D (on lead, grounds only) ♣ ⛩ ★ ◆ 🚶 (by appt) 🚶

West Yorkshire

Bolling Hall, Bowling Hall Rd, Bradford, West Yorkshire (tel Bradford [0274] 723057). 1½ m SE of town centre on A650. SE 1831 (OS 104). 15th-century manor house and tower, extended in the 17th and 18th centuries. The house is now a museum containing period furniture (including several pieces by Chippendale) and displays of local life. Open T-Su and Bank Hol M Apr to Sept 1000-1800; Oct to Mar 1000-1700; closed 25, 26 Dec, 1 Jan, Good Fri. ⊖ P WC ♿ (limited access) ⊟ (by appt) ♣ ⛩ ★ ◆ ✱ ● (with permission, no flash)

Bramham Park, Wetherby, West Yorkshire (tel Boston Spa [0937] 844265). 7 m S of Wetherby on A1, turn W. SE 4041 (OS 105). Queen Anne mansion built by Robert Benson in the Italian manner. The interior was restored in an authentic style in the early 20th century after a fire in 1828. The park retains its original Classical layout, incorporated excellent vistas inspired by Versailles. Open mid-June to end Aug Su, T-Th and Bank Hol M 1315-1730. ⊖ P WC ♿ (limited access) ⊟ (by appt) D (on lead, grounds only) ☕ (for parties only) ⛩ ✱ 🚶 (by appt for parties) ● (not in house)

East Riddlesden Hall, near Keighley, West Yorkshire (tel Keighley [0535] 607075). 2 m NE of Keighley on A650. SE 0742 (OS 104). 17th-century manor house, with a fine banqueting hall, plasterwork, panelling and a Restoration-period staircase. The grounds contain a timber-framed tithe barn, considered one of the best in the country, with a collection of farm implements. Open Apr to Nov S, Su pm; May to Oct W-F pm; Bank Hol M pm; open Easter week (exc Good Fri) pm. ⊖ P WC ♿ (limited access) ⊟ ♣ ☕ ◆ ✱ 🚶 (by appt) ● (no tripods, flash) NT

Lotherton Hall, Aberford, West Yorkshire (tel Aberford [0532] 813259). 10 m E of Leeds, on B1217. SE 4436 (OS 105). Edwardian house created in Neo-Classical style around an 18th-century building. It is now a museum, containing a collection of 18th-century paintings, Chinese ceramics and 19th- and 20th-century furniture, including modern works. There is also a costume gallery, and a collection of silver race cups. The gardens are attractive. Open T-Su and Bank Hol M, am and pm. ⊖ 🅿 WC 🔽 🚻 (by appt) D (not in house) ♣ 🍽 𝄡 (by appt)

Manor House, Castle Yard, Ilkley, West Yorkshire (tel Ilkley [0943] 600066). In town centre. SE 1147 (OS 104). Elizabethan manor house built on the site of a Roman fort, with some of the Roman foundations visible. There is a display of Roman materials, and a programme of exhibitions by local artists and craftsmen. Open T-Su and Bank Hol M am and pm. Closed Christmas and New Year. ⊖ 🅿 WC 🚻 (by appt) ★ 𝄡 (by appt) ●

Nostell Priory, Wakefield, West Yorkshire (tel Wakefield [0924] 863892). 8 m SE of Wakefield on A638, turn N. SE 4017 (OS 111). Palladian house built on the site of an old abbey by the young James Paine in the 1730s, and extended by Robert Adam in 1766. The state rooms are all by Adam and include fine plasterwork, and furniture made especially for the house by Thomas Chippendale. There is a large park, with fine trees, a museum of vintage and veteran motor cycles, and an aviation museum.Open July and Aug daily exc F pm; Mar to June, Sept and Oct Bank Hol M, T pm; S and Su pm. ⊖ 🅿 WC 🔽 (limited access) 🚻 ♣ 🍽 ♦ ❄ 𝄡 (by appt) ● (no tripods, flash) ⊀ NT

Oakwell Hall, Birstall, West Yorkshire (tel Batley [0924] 474926). 6 m SW of Leeds, off A652. SE 2226 (OS 104). Moated manor house dated 1583. The house retains its 17th-century decoration and oak furniture. The house features in Charlotte Brontë's novel *Shirley*. Open daily am and pm (Su pm only). ⊖ 🅿 WC 🔽 (limited access) 🚻 (by appt) D (not in house) ♣ 🍽 (limited opening) ⌂ ★ ♦ ❄ 𝄡 ● (permission required) ⊀ ⋏

Red House, Gomersal, West Yorkshire (tel Cleckheaton [0274] 872165). 7 m SE of Bradford on A651. SE 2026 (OS 104). Red-brick house of the Restoration period, often visited by Charlotte Bronte who described it in *Shirley*. It is furnished in the style of that period. Open daily am and pm (Su pm only). ⊖ 🅿 WC 🚻 ♣ ⌂ ★ ♦ ❄ 𝄡 (by appt) ●

Shibden Hall, Halifax, West Yorkshire (tel Halifax [0422] 59454 ext 235). 1 m E of town centre, on A58. SE 1025 (OS 104). An early 15th-century timber-framed house with later additions. The interior contains period rooms of the 16th-18th centuries. The 17th-century barn and outbuildings contain the West Yorkshire Folk Museum, with a collection of old agricultural implements and craft workshops. Open March to Nov daily am and pm (Su pm only). ⊖ 🅿 WC 🔽 (limited access) 🚻 (by appt) D ♣ 🍽 ⌂ ♦ ❄ 𝄡 (for parties, by appt) ⊀

Scotland

Wick

WESTERN
ISLES

HIGHLAND

Brodie Castle

Inverness

GRAMPIAN

Leith Hall
Haddo
House

Fort William

Provost
Skene's
House
Aberdeen

Balmoral Castle

TAYSIDE

Torosay Castle

Dundee

Perth
Scone Palace

Hill of
Tarvit

Falkland Palace

CENTRAL

Hill
House

Stirling

FIFE

Culross Palace

2
3
Tenement
House
1
Edinburgh
4
7 5 6
8
Lennoxlove

Glasgow
LOTHIAN
Winton House

Pollok House

Manderston

STRATHCLYDE

Mellerstain

Traquair
House
Abbotsford
House

Bowhill
BORDERS

Ayr

Hawick

Penkill
Castle

NORTHUMBER-
LAND

DUMFRIES
Maxwelton
House
&
Carlyle's
Birthplace

Dumfries
Newcastle-
upon-Tyne

GALLOWAY
Rammerscales

Carlisle

CUMBRIA
DURHAM

1 House of the Binns
2 Hopetoun House
3 Dalmeny House
4 Georgian House
5 John Knox House Museum
6 Holyroodhouse
7 Gladstone's Land
8 Gosford House

Abbotsford House

Melrose, Borders

Sir Walter Scott's home by the River Tweed could be called the original country house open to the public, for even before the novelist's death in 1832 it had become a target for tourists, and it was officially opened to visitors the following year. The king of the romantic historical movement in English literature created an 'antient' property where previously there had been a modest farmstead, and invented a new name for it invoking a far-off monastic tenure of the Abbey of Melrose. To love Scott is to love Abbotsford and its medieval-magpie interior. The place was rebuilt by him in 1822-24 with the enthusiasm of an amateur, and with the help of an undistinguished architect, and it reflects Scott and his interests in an entirely personal way. The exterior design is a Tudor/Scottish-traditional 'mixed salad', with big sash windows to let in plenty of light. Inside, the light is considerably dampened by the massive, dark-coloured panelling and furnishings, and the crowding together of relics, memorabilia and solid furniture. Scott had bought the farm in 1811, and it was here that he began to write the series of 'Waverley' novels in 1814. After the completion of Abbotsford in 1824, all of his work was written in his personally designed study on the ground floor. He enjoyed his fame, and the library (large bay windows giving on to the Tweed) and entrance hall (stained glass) are on a scale to accommodate that hospitality for large numbers of friends and admirers for which Scott was renowned.

☎ Galashiels (0896) 2043

2 m SE of Galashiels on A7 (which becomes A6091); turn W onto B6360

NT 5034 (OS 73)

Open mid Mar to end Oct: M-S 1000-1700; Su 1400-1700; reduced rates for parties

⊖ P WC ♿ ⊟ (by appt) ♣ ☕ ⛩ ◆ ✿
(by appt for parties)

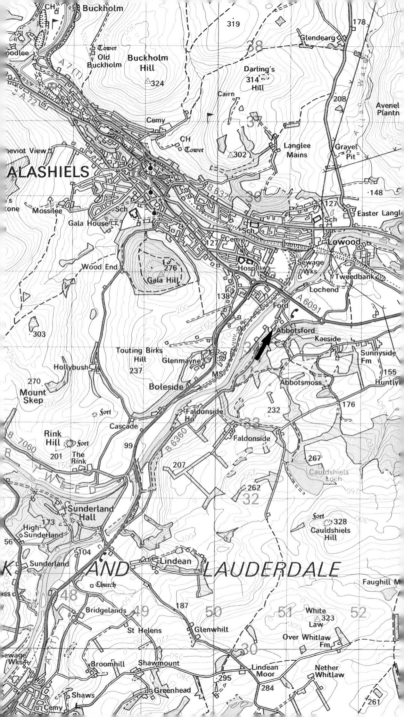

Manderston

Duns, Borders

Manderston, a lavish Edwardian country house, is lived in by the descendants of James Miller, who largely created it in the 1900s. The first house, of the 1790s, was built by Dalhousie Weatherstone, and the estate was bought in 1864 by William Miller, who had made a fortune from trading in hemp and herrings. In 1893 his son Sir James married Lady Miller, daughter of Lord Scarsdale. The latter was the owner of Kedleston Hall in Derbyshire, one of Robert Adam's great achievements, and it was probably to impress his new father-in-law that Miller at once began to plan a new house entirely in the Adam style. His architect was John Kinross, who was given *carte blanche* in the matter of finance, an architect's dream come true. The stables were built first, in 1895, with the grooms' rooms panelled in mahogany and the horses' stalls in teak with marble plaques for their names. The house itself was given a new wing, a new entrance front and a service court. The interior, with its string of reception rooms designed for formal entertaining, is pure Adam, with one of the chimneypieces and the ballroom ceiling actually copied from Kedleston. The showpiece is the drawing room leading to the ballroom, with furniture in the Louis XVI style, white silk curtains and walls hung with silk brocade. Almost as much care was lavished on the servants' quarters: the basement, which runs the entire length of the house, is unchanged, and provides an interesting example of the sophisticated domestic arrangements of such a household.

☎ Duns (0361) 83450

1½ m E of Duns on A6105 continue on minor road to Buxley

NT 8154 (OS 67)

Open mid May to end Sept Th, Su and Eng Bank Hol M 1400-1730; other times for parties by appt

⊖ 🅿 WC ♿ (limited access) 🚉 (by appt)
D (grounds) ◆ ♣ 🍴 ☗ ✂ 🏇 ● ⛰ 🚶

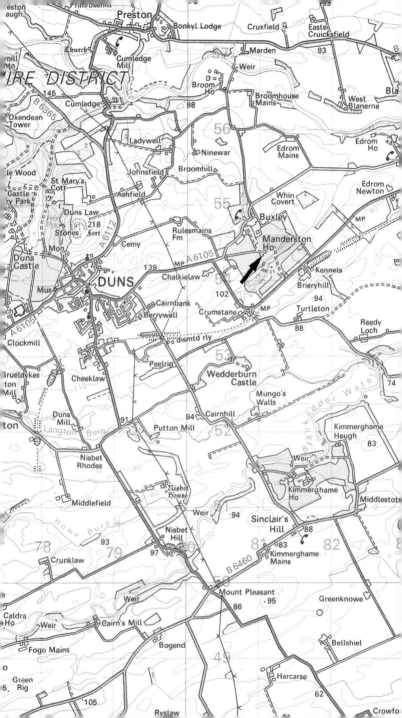

Mellerstain

Gordon, Borders

The foundations for a new mansion at Mellerstain were laid in 1725 for the owner, Lady Grisell Baillie, who had intended to build a modest Palladian house on the site of the old building. However, only the simple square wings, designed by William Adam, were built, and when Lady Grisell's nephew George Baillie inherited in 1759, he abandoned her project and called in Robert Adam, William's much more famous son, to complete the house. The exterior of the house is in the castle style which was then becoming popular, and which Adam had used at Culzean and elsewhere. The interiors, in Adam's most assured light Classical style, are a delight, and are still complete with their original ceilings, friezes and fireplaces. The library (completed in 1773), with its white carved-wood bookcases, mahogany doors and green-and-white marble fireplace, is one of the finest rooms Adam designed, and the dining room, music room and drawing room, all decorated in their original colours, show the highest degree of skill and craftsmanship. The long gallery, which is used to display a collection of 18th- and 19th-century costume, would certainly have been as fine as the library, but the long barrel-vaulted ceiling was never completed. The house contains some good 18th-century furniture, though Adam did not design it himself as he did in many cases, and there is also an exceptional collection of paintings, Italian religious subjects as well as works by Van Dyck, Gainsborough, Ramsay, Jacob Ruysdael and Constable.

☎ Gordon (057 381) 225

8 m NW of Kelso on A6089 turn W to Mellerstain

NT 6439 (OS 74)

Open Easter weekend and May to end Sept daily exc S 1230-1700; reduced rates for parties by appt

WC ♿ (limited access) ♨ (by appt) D (grounds only) ♠ ▬ ◆ ✳ 🍴 (parties only) ●

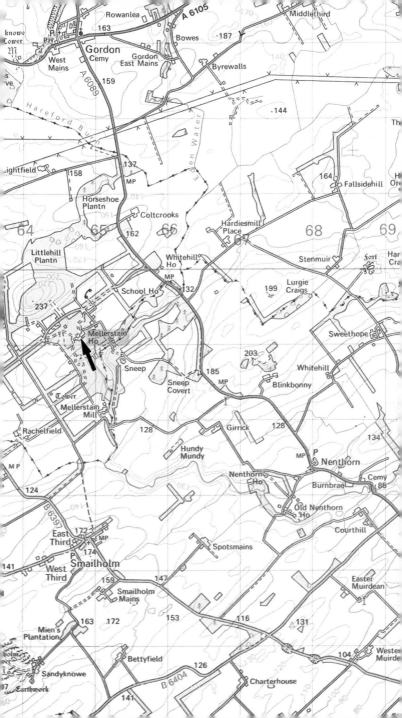

Traquair House

Innerleithen, Borders

The site of Traquair, in the upper valley of the Tweed, was known as a royal hunting lodge for the kings of Scotland until James III gave away the estate in 1478. It was soon bought by the Earl of Buchan, who gave it to his son, James Stuart. He became the 1st laird, and his descendants still live there. The small fortress built by James Stuart – a strong, battlemented tower – forms part of the present house, which his son and grandson began to build around it. Other substantial additions followed in the 17th century, notably the two wings forming a forecourt at the front, and the terrace with its two domed pavilions overlooking the river. The house was extended modestly; the Stuarts were never rich or powerful enough to refashion in a grand style. The main façade reveals a building quite unlike any Tudor or Jacobean house in England; it is perhaps more reminiscent of a French château. The low storeys, small windows and steep roofs, and the lack of embellishments all give an impression of austerity. The Stuarts were often persecuted for their Catholic and Jacobite sympathies, and the house contains many relics of the family's loyalty to lost causes. An oak carving of the Royal Arms of Scotland commemorates the visit of Mary Queen of Scots, whose rosary and crucifix are also shown. Miniatures of the exiled royal Stuarts and a fine collection of Jacobite glass mark the active support of later Stuarts, by now earls of Traquair. There are other treasures: historic documents, china, needlework, and the 18th-century library, where the books occupy their original positions.

☎ Innerleithen (0896) 830323

1 m S of Innerleithen on B709

NT 3235 (OS 73)

Open Easter to mid Oct daily 1330-1730 (July to mid Sept 1030-1730); parties at other times by appt

⊖ (1 m) P WC 🅰 (limited access) ⊟ D (grounds only) ♣ ☕ ⊓ ◆ ﹅ ● ⚇ ⚇

Falkland Palace

Falkland, Fife

Falkland passed to the Stuarts in 1370, and had become a favourite place for hunting and relaxation by 1500 when James IV began to build a new palace to the south of the old – probably 13th-century – castle. This building consisted of three oblong blocks ranged round a central courtyard, and the style is Scottish Gothic, but the additions and alterations made by James V between 1537 and 1542 are entirely different. His courtyard façade, far more advanced architecturally than anything else in Scotland or England at the time, is pure French Renaissance, reminiscent of the châteaux of Blois or Chambord. The reason for this surprising piece of building was James' visit to the French court in 1537 and his subsequent return with a French bride, Madeleine, daughter of François I. The architect was Sir James Hamilton of Finnart, who had himself spent several years at the French court, and all the craftsmen were either French or trained in France. The angels decorating the buttresses on the south range were carved in 1539 by 'Peter the Flemishman', and the medallions with busts on the east range, which local tradition claims as portraits of personages connected with the palace, were done in 1538 by a French craftsman, Nicholas Roy. Sadly, the only interior to survive of these magnificent state rooms is the Chapel Royal, and even here only the elegant screen is of this period; the rest of the decorations date from 1635, when the palace was redecorated for Charles I's visit.

☎ Falkland (0337) 57397

16 m SE of Perth on A912 at Falkland

NO 2507 (OS 59)

Open Apr to end Sept M-S 1000-1800, Su 1400-1800; Oct S 1000-1800, Su 1400-1800; parties at other times by appt

⊖ 🅿 WC 🍴 ♿ ⌙ ◆ ⚹ 🍴 ● (no flash)

Holyroodhouse

Edinburgh, Lothian

Holyroodhouse has a long history, the name itself being the legacy of the 11th-century queen who was to become St Margaret of Scotland, who died here, bequeathing her precious relic of Christ's cross, the 'holy rood', to her son David. The splendid 17th-century palace we see today grew from the guesthouse attached to a medieval monastery, and was begun under James IV in 1501. After the Restoration of 1660 Charles II decided to rebuild the palace, though he never came to see the result, and the great period of building took place from 1671 to 1680 at his behest, under the supervision of Sir William Bruce. The oldest rooms are those in the James IV tower, home of Mary Queen of Scots, where Lord Darnley plotted and David Rizzio met his death. The Queen's inner chamber has a lovely oak coffered ceiling with 17th-century painted panels, while the outer chamber has a 17th-century chimneypiece and a timber ceiling commemorating Mary's marriage to the Dauphin François. The gallery, which extends along the whole northern range and joins the old apartments to the 17th-century wing, has portraits of eighty-nine Scottish monarchs set into the panelling. These were commissioned by Charles II, and painted by the Dutch artist Jacob de Witt, who worked on them at the palace for two years. The finest room of all is the privy chamber (now called the morning drawing room as Queen Victoria used it for this purpose) where the splendid plaster ceiling, intricately carved woodwork, and wallpaintings are all original.

☎ Edinburgh (031) 556 7371

In centre of Edinburgh at E end of Canongate

NT 2773 (OS 66)

Open mid Mar to late Oct M-S 0930-1715, Su 1030-1630; late Oct to mid Mar M-S 0930-1545

⊖ F WC ⟨⟩ ⊟ ⟨⟩ ◆ ⟨⟩ ✗ (compulsory)
● (not in house)

Hopetoun House

South Queensferry, Lothian

On the southern shore of the Firth of Forth, standing in its beautiful park, Hopetoun House is something of a surprise. Looking like an English country mansion, it was designed and built by Scottish architects and craftsmen for a family of French origin who had been settled in Scotland for more than 150 years. Completed in 1703, the house was planned as a central block with two small wings, of which the central block remains. Some fifteen years later the Earl of Hopetoun commissioned William Adam to improve and enlarge the house. The east front was refaced, he added the bays projecting from it, built concave colonnades on either side and the large wings, each enhanced by a tower with a cupola. His sons Robert and John put the final touches to the house after his death, and decorated the interior. Much of the original Adam decoration survives – in the hall and in the yellow and red drawing rooms, which are important examples of early Adam work. Only the furnishings are by an Englishman, James Cullen of London; they were all made specially for Hopetoun and show the quality of Cullen's work. There is also a rich collection of paintings. Some of the earlier rooms also survive, while the state dining room is late Regency, created in about 1820 by the 4th Earl. With its elaborate curtains, over-decorated cornice and gold cloth on the walls, it makes an interesting comparison with the furnishings of the other rooms shown to the public.

☎ Edinburgh (031) 331 2451

3 m W of Queensferry on A904 turn to Abercorn at junction with B8020

NT 0979 (OS 65)

Open Easter weekend, then May to mid-Sept daily 1100-1730; parties by appt at other times

⊖ (limited) P WC ⬛ (limited access) ⊟ (by appt) D (grounds) ♣ 🍴 ⊼ ◆ 🎪 ⚹ ⚘ ● ⚘

Winton House

Pencaitland, East Lothian, Lothian

Winton House, secondary residence of the important Seton family, saw three main periods of building, the late 15th, the 17th and the early 19th centuries. The first house was built about 1480, and though it was later 'burnt by the English', the barrel-vaulted ground floor and lower part of the walls survived and are incorporated into the present house. The house may have been restored about 1600, but the lovely tall, gabled, Jacobean central part was built in 1620 by William Wallace, the King's Master Mason. A particular feature of this phase of building is the extraordinary carved and sculptured chimneys, very unusual as they are of stone rather than brick. The third building programme began in 1800, when Colonel Hamilton enlarged the house, building a new north (entrance) front in the 'Gothick' style, together with battlemented wings on the north and west. The entrance hall leads into the octagon hall, built above the original open courtyard and giving a good view of the carved window frames. An ante-room then leads into the east wing, which is part of the old house though the ceilings are mainly of the Jacobean period. The finest of these is in the library. The drawing room, originally the great hall, has a fine Renaissance chimneypiece and an ornate plaster ceiling with the arms of the Setons. A turret stair leads from the first-floor bedrooms in the tower to the top, which provides a fine close-up view of the chimneys, while six storeys below is the oldest room of all, the barrel-vaulted kitchen, now used as the dining room.

☎ Pencaitland (0875) 340 222

7½ m SW of Haddington on A6093 turn N on B6355 for 1 m then S

NT 4369 (OS 66)

Open throughout year by written appt

⊖ (1 m walk) 🅿 WC ☒ (by appt) 🚌 ♣

🍽 (by appt) ⚘ 🕱 (compulsory)

Torosay Castle

Craignure, Isle of Mull, Strathclyde

The parish of Torosay was acquired in the early 19th century by Colonel Campbell of Possil, who demolished the existing Queen Anne house and commissioned David Bryce, the Edinburgh architect who pioneered the Scottish baronial style, to build a much larger house on the same site. The 'castle' was completed in 1858, and was then called Duart House. The diary of a guest tells us the family were well pleased with it, and throughout the Victorian and Edwardian periods they entertained lavishly, well-known visitors to the house including Dame Nellie Melba and Lillie Langtry. In 1897 it was left to Walter Murray Guthrie, grandfather of the present owner, who put it on the market briefly before he visited it, then fell in love with it and withdrew it from sale. It was he who commissioned Sir Robert Lorimer to design and lay out the three Italian terraces and statue walk that now connect the house to the old walled garden. Torosay Castle has recently been restored and the interior reflects its heyday of prosperity. The front hall is hung with rows of red deer antlers, and there is a painting by Landseer at the top of the stairs; the main hall is dominated by a fine portrait by Poynter of Olive Guthrie, who lived here until 1945, and there are more family portraits in the other rooms, including the dining room, where there is also a painting by Sir John Leslie, Olive Guthrie's father, painted in 1855 and hung in the Royal Academy that year. The boudoir contains exhibits on the theme of the Loch Ness Monster.

☎ Craignure (068 02) 421

On Island of Mull, W of Duart Point on A849

NM 7235 (OS 49)

Open May to end Sept daily 1030-1730, Apr and Oct by appt

⊖ P WC ♿ (limited access) ♨ (by appt) D (on lead, grounds only) ♣ ♥ ⊼ ◆ ☼

Scone Palace

Perth, Tayside

Scone is the centre of Scotland both geographically and historically: the first recorded Scottish Parliaments were here, and here was kept the famous Stone of Scone upon which the kings of Scotland were crowned. There was an ancient religious foundation at Scone; its ruins formed the base of the house built in 1580 by the Ruthven family, Earls of Gowrie. After 1600 James VI gave the property to Sir David Murray, and it has remained in the family ever since. The present house, built between 1802 and 1813 for the 3rd Earl of Mansfield, was designed by William Atkinson, the pupil of James Wyatt. Built of red sandstone in a sober Gothic style, it is similar to Wyatt's earlier buildings. The interior is entirely consistent with the exterior, the hall being vaulted or beamed, with every detail in the Gothic manner. The palace contains an amazingly rich variety of objects and works of art, among the earliest being a 16th-century needlework panel and pieces of a set of embroidered bed-hangings worked by Mary Queen of Scots and her ladies. The dining room has a unique collection of European ivories and two carved ivory mirrors; the charming little ante-room, painted in white, silver and gold, contains Chinese vases and Chinese Chippendale chairs, but the most spectacular furniture is in the drawing room, where there are 18th-century *boulle* commodes, tables and clocks, a set of tapestry-covered Louis XV chairs and an exquisite marquetry writing-table made for Marie-Antoinette by Reisener and stamped with her own cipher.

☎ Scone (0738) 52300

3 m N of Perth on A93, turn SW at Scone

NO 1026 (OS 53)

Open Good Fri to mid Oct M-S 0930-1700, Su 1330-1700 (July, Aug Su opening 1000);

⊖ (limited, not Su) P WC 🚻 🚽 (by appt) D (on lead, grounds only) 🍴 🍷 🎏 ◆ 🎿 🎎 🎎

Borders

Bowhill, Selkirk, Borders Region (tel Selkirk [0750] 20732). 3 m W of Selkirk on A708, turn S. NT 4227 (OS 73). 18th- and 19th-century mansion, the home of the Scotts of Buccleuch. The house contains a famous art collection, including works by Leonardo, Guardi, Canaletto, Claude and Gainsborough, and many *objets d'art*. There are also relics of the Duke of Monmouth, Queen Victoria and Sir Walter Scott, and a collection of 16th- and 17th-century English miniature portraits. House open early July to mid Aug 1300-1600, Su 1400-1800; grounds open May to Aug 1300-1630, Su 1400-1800. Parties at other times (May to Aug) by appt. ▐ WC ♿ ⊟ (by appt) D (on lead, grounds only) ✿ 🐾 ⋔ ◆ ✲ ✗ (by appt) ⚹ ⚘

Dumfries & Galloway

Carlyle's Birthplace, Ecclefechan, Dumfries & Galloway Region (tel Ecclefechan [057 63] 666). 7 m SE of Lockerbie on A74. NY 1974 (OS 85). 18th-century artisan's house, built by Thomas Carlyle's father and uncle, in which the writer was born in 1795. There is a collection of mementos and personal relics. Open Easter to end Oct daily (exc Su) 1000-1800. ⊖ ▐ ⊟ (by appt, max 20)

Maxwelton House, near Moniaive, Dumfries & Galloway Region (tel Moniaive [084 82] 385). 13 m NW of Dumfries on B729. NX 8289 (OS 78). Early 17th-century house, incorporating an earlier tower; a stronghold of the Earls of Glencairn. It was the birthplace in 1682 of Annie Laurie, immortalised in song, and there is a museum of agricultural and early domestic life. House open July to Aug M-Th 1400-1700; grounds open Apr to Sept M-Th 1400-1700. ⊖ (limited) ▐ WC ♿ (limited access) ⊟ (by appt)

D (on lead, grounds only) ✿ 🐾 (parties only, by appt) ✲ ✗ (compulsory) ● (permission required) ⚹ ⚘

Rammerscales, Lockerbie, Dumfries & Galloway Region (tel Lochmaben [038 781] 361). 5 m W of Lockerbie, on B7020. NY 0877 (OS 85). Manor house built in 1760 for Dr James Mounsey. There is a fine circular staircase and elegant rooms, with a collection of relics of the Jacobite movement and of Flora Macdonald. Open end July to early Sept on certain days (phone for details). ▐ WC ⊟ (by appt) ✿ ⋔ ✗ (by appt, parties only)

Fife

Culross Palace, Culross, Fife Region. 7 m W of Dunfermline, off A985. NS 9885 (OS 65). Royal palace built in the early 17th century on the site of an earlier palace and monastery. It had been the home of many medieval monarchs, and it was the birthplace of Charles I. There are fine painted wooden panels, and terraced gardens. Open daily am and pm (Su pm only). ⊖ ▐ WC ⊟ ✿ ⋔ ◆ ✲ ✗ ● (no flash)

Hill of Tarvit, near Cupar, Fife Region (tel Cupar [0334] 53127). 2½ m SW of Cupar on A916, turn E. NO 3711 (OS 59). Old mansion house, heavily rebuilt in the early 20th century. It has a fine collection of furniture, tapestries, porcelain and paintings. House open May to end Sept daily (1400-1800; Oct S and Su only 1400-1800. Gardens open daily 1000-dusk. ▐ WC ♿ (limited access) ⊟ (by appt) D (on lead, grounds only) ✿ 🐾 ⋔ ✲ ● ⚘ NTS

Grampian

Balmoral Castle, near Ballater, Grampian Region (tel Crathie [033 84] 334). 8 m SW of Ballater on A93, turn S. NO 2595 (OS 44). Mid 19th-century castle in Baronial style, built for Queen Victoria and Prince Albert. The castle is the Highland residence of Her Majesty the Queen, and is set in fine forested grounds. The ballroom contains an exhibition of paintings and works of art. Open May to July daily (exc Su) 1000-1700. ⊖ (1000 yds) 🅿WC 🅱 🖩 (by appt) D (on lead, grounds only) ♠ 🍺 ◆ ✻ ● (not in castle) 🕭

Brodie Castle, near Nairn, Grampian Region (tel Brodie [030 94] 371). 6 m E of Nairn off A96. NH 9757 (OS 27). 16th- and 17th-century castle rebuilt after being burned in 1645. It contains a fine collection of paintings, including Dutch Old Masters, Impressionists and others. House open mid-Apr to end Sept daily am and pm; grounds open all year daily am and pm. ⊖ 🅿WC 🖩 (limited access) 🅱 D (grounds only) ♠ 🍺 (limited) ⊓ ◆ NTS

Haddo House, near Methlick, Grampian Region (tel Tarves [065 15] 440). 24 m N of Aberdeen on B9005. NJ 8634 (OS 30). Classical house built in the 1730s by William Adam for the Gordons of Haddo. The house has an elegant front, with a divided, sweeping stair. The interior was restyled in the later 19th century. There is a fine collection of portraits, including those of many mid 19th-century statesmen; the chapel has stained glass by Burne-Jones. House open May to end Sept daily 1400-1800 (exc 2 days – F,S – in mid May); grounds open all year daily 0930-2030. ⊖ 🅿 WC 🖩 🅱 (by appt) D (on lead, grounds only) ♠ 🍺 ⊓ ◆ ✻ 🕭 (compulsory exc Su) ● (no flash) 🕮 🕭 NTS

Leith Hall, Kennethmont, Grampian Region (tel Kennethmont [046 43] 216). 34 m NW of Aberdeen, on B9002. NJ 5429 (OS 37). Mansion begun in the 1650s, with wings added in the 18th and 19th centuries; the home of the Leith

family. The house contains family relics. The garden is pleasant, with nature walks in the grounds. House open May to end Sept daily 1400-1800; grounds open daily all year sunrise to sunset. ⊖ 🅿WC 🖩 (limited access) 🅱 (by appt) D (on lead, grounds only) ♠ ⊓ ◆ ✻ 🕭 (compulsory) ● (no flash) 🕭 NTS

Provost Skene's House, Aberdeen, Grampian Region (tel Aberdeen [0224] 641086). In Guestrow, off Broad St, in city centre. NJ 9305 (OS 38). 17th-century house, furnished with period rooms and displays of local history and domestic life. Open all year M-S am and pm. ⊖ 🅿WC 🅱 (by appt) 🍺 ★ ◆ ✻

Lothian

Dalmeny House, South Queensferry, Lothian Region (tel 031-331 1888). 7 m W of Edinburgh off A90. NT 1678 (OS 65). The first Gothic Revival house in Scotland, designed in 1814 by William Wilkins. The house contains collections of French furniture and porcelain, the Earl of Rosebery's paintings, tapestries and a Napoleon room. Open early May to end Sept daily (exc F, S) 1400-1730; other times by appt. ⊖ (1 mile) 🅿WC 🅱 (by appt) D (on lead, grounds only) ♠ 🍺 ✻ 🕭 (compulsory) ● (permission required)

Georgian House, 7 Charlotte Square, Edinburgh, Lothian Region (tel 031-225 2160). In city centre. NT 2574 (OS 66). One of the houses of Robert Adam's designs for Charlotte Square, praised as his masterpiece of urban architecture. The house has been restored as a typical 18th-century mansion. Open Apr to end Oct M-S am and pm, Su pm only; Nov S am and pm, Su pm only. ⊖ 🅱 ◆ ✻ 🕭 ●

Gladstone's Land, Lawnmarket, Edinburgh, Lothian Region (tel 031-226 5856). In city centre. NT 2573 (OS 66). House built in 1620 and furnished as a typical 'Old Town' house with remarkable painted wooden ceilings. It is named after Thomas Gledstanes, who acquired the house shortly after it was built. The ground floor contains a shop front and goods of the period. Open Easter to end Oct M-S am and pm, Su pm only; Nov, S am and pm, Su pm only. ⊖ 🏠 ◆ ✴ ✗ ● NTS

Gosford House, near Longniddry, Lothian (tel Aberlady [08757] 201). 14 m E of Edinburgh, off A198. NT 4578 (OS 66). A Neo-Classical house built by Robert Adam and William Young. Open June and July W, S and Su 1400-1700. ⊖ 🅿 ♿ 🏠 (by appt) D (on lead, grounds only) ♣ ◆ ✴ ● (no flash) ♪

House of the Binns, Linlithgow, Lothian Region (tel Philipstoun [050 683] 4255). 3½ m E of Linlithgow on A904. NT 0578 (OS 66). The house was begun in the 15th century, then remodelled in the 17th in transitional style, and modified again in the 19th century. There are fine moulded ceilings dating from 1630, and a 19th-century folly in the grounds. Open Easter, May to Sept daily (exc F) pm. ⊖ 🅿 🏠 ♣ NTS

John Knox House Museum, 45 High St, Edinburgh, Lothian Region (tel 031-556 6961). In city centre. NT 2573 (OS 66). 15th- and 16th-century town house, probably the home of religious reformer John Knox from 1561 to 1572. The house has fine wooden galleries of the late 16th century, and a painted ceiling of the early 17th century. There is a collection of items relating to Knox, and pictures of the old Edinburgh. Open M-S 1000-1300 (closed 25th Dec to 3rd Jan). ⊖ 🅿 (limited) 🏠 (by appt) ◆ ● (no video)

Lennoxlove, Haddington, Lothian Region (tel Haddington [062 082] 3720). 1½ m SE of Had-dington on A6137, turn E. NT 5172 (OS 66). Fortified tower house first erected in the mid-14th century and reworked by the Duke of Lauderdale in the reign of Charles II. There is fine 17th- and 18th-century furniture, 18th-century porcelain, and many notable portraits. Open Apr to end Sept W, S, Su 1400-1700; other times by appt. ⊖ (½ mile) 🅿 WC 🏠 (by appt) D (on lead, grounds only) ♣ ● 🏠 ◆ ✴ ✗ ♪

Strathclyde

Hill House, Upper Colquhoun St, Helensburgh, Strathclyde Region (tel Helensburgh [0436] 3900). In town centre. NS 2982 (OS 56). Built in 1902 for Walter Blackie, one of the finest domestic houses built by the Glasgow Art Nouveau architect Charles Rennie Mackintosh. Its furnishings were all designed by Mackintosh; the house overlooks the River Clyde. Open daily pm. 🅿 WC 🏠 ♣ 🏠 ◆ ✴ ● (no flash, tripods, video) NTS

Penkill Castle, Girvan, Strathclyde Region (tel Old Dailly [046 587] 261). 2½ m E of Girvan off A77. NX 2398 (OS 76). 16th-century tower house, rebuilt in the 19th century, and the home in the 1860s of painter William Bell Scott. There is a collection of Pre-Raphaelite paintings. The house has special connections with the Rossettis. Open daily by appt. 🅿 WC 🏠 ♣ ● (inc with tour) ★ 🏵 ✗ (compulsory) ♪

Pollok House, Pollok Park, Glasgow, Strathclyde Region (tel 041-632 0274). 3½ m S of city centre. NS 5561 (OS 64). Mid-Georgian house extended in the early 19th century, and containing the Stirling Maxwell collection of Spanish and other European paintings, with works by El Greco, Murillo, Goya, Signorelli and Blake. Open daily 1000-1700 (Su 1400-1700 only); closed 25th Dec, 1st Jan. ⊖ 🅿 WC ♿ 🏠 (by appt) ♣ ● 🏠 ★ ◆ ✴ ✗ (by appt) ● (permission required) ♪

Index